CYTOLOGY OF
NON-GYNAECOLOGICAL SITES

INTERNATIONAL HISTOLOGICAL CLASSIFICATION OF TUMOURS

No. 17

CYTOLOGY OF NON-GYNAECOLOGICAL SITES

G. RIOTTON

Head, WHO Collaborating Centre for Nomenclature in Cytology, Professor and Director, Centre de Cytologie et de Dépistage du Cancer, Geneva, Switzerland

W. M. CHRISTOPHERSON

Professor and Chairman, Department of Pathology, University of Louisville School of Medicine, Louisville, Kentucky, USA

in collaboration with

R. LUNT

Scientist, World Health Organization, Geneva, Switzerland

and cytologists in 6 countries

WORLD HEALTH ORGANIZATION

GENEVA

1977

ISBN 92 4 176017 6

© World Health Organization 1977

Publications of the World Health Organization enjoy copyright protection in accordance with the provisions of Protocol 2 of the Universal Copyright Convention. For rights of reproduction or translation of WHO publications, in part or *in toto*, application should be made to the Office of Publications, World Health Organization, Geneva, Switzerland. The World Health Organization welcomes such applications.

The designations employed and the presentation of the material in this publication do not imply the expression of any opinion whatsoever on the part of the Director-General of the World Health Organization concerning the legal status of any country, territory, city or area or of its authorities, or concerning the delimitation of its frontiers or boundaries.

The mention of specific companies or of certain manufacturers' products does not imply that they are endorsed or recommended by the World Health Organization in preference to others of a similar nature that are not mentioned. Errors and omissions excepted, the names of proprietary products are distinguished by initial capital letters.

Authors alone are responsible for views expressed in this publication.

PRINTED IN SWITZERLAND

LIST OF PARTICIPANTS

WHO International Reference Centre for Nomenclature in Cytology [1]

Head of Centre

DR G. RIOTTON, Centre de Cytologie et de Dépistage du Cancer, Geneva, Switzerland

Participants

DR W. M. CHRISTOPHERSON, Department of Pathology, University of Louisville, Kentucky, USA

DR S. HATTORI, Laboratory of Diagnostic Cytology, Centre for Adult Diseases, Osaka, Japan

DR O. A. N. HUSAIN, St Stephen's Hospital, London, United Kingdom

DR L. KOSS, Department of Pathology, Montefiore Hospital, Bronx, New York, USA

DR E. MCGREW, Department of Pathology, University of Illinois School of Medicine, Chicago, Illinois, USA

DR M. NASIELL, Department of Cytology, Sabbatsberg Hospital, Stockholm, Sweden

DR V. NUOVO, Centre de Cytopathologie, Monte Carlo, Principauté de Monaco

DR A. C. PETROVA, Cytology Laboratory, Cancer Research Centre of the USSR Academy of Medical Sciences, Moscow, USSR

DR R. O. K. SCHADE, The General Hospital, Newscastle upon Tyne, United Kingdom

DR S. SHIDA, Department of Surgery, Dokkyo University School of Medicine, Mibu-Machi, Shimotsuga-Gun, Tochigi, Japan

DR L. WOOLNER, Mayo Clinic, Rochester, Minnesota, USA

DR J. ZAJICEK, Department of Clinical Cytology, Karolinska Hospital, Stockholm, Sweden

ACKNOWLEDGEMENT

The WHO Collaborating Centre for Nomenclature in Cytology wishes to express its gratitude to Miss Valerie Rawyler, Chief Cytotechnologist, Centre de Cytologie et de Dépistage du Cancer, Geneva, for her invaluable help in the preparation of the material for this publication.

---

[1] Following a decision taken by the World Health Organization in 1974 in the interest of uniformity, all WHO-designated centres have been re-named WHO Collaborating Centres; thus the above centre is now known as the WHO Collaborating Centre for Nomenclature in Cytology.

ALREADY PUBLISHED IN THIS SERIES:

No. 1. Histological typing of lung tumours (1967)

No. 2. Histological typing of breast tumours (1968)

No. 3. Histological typing of soft tissue tumours (1969)

No. 4. Histological typing of oral and oropharyngeal tumours (1971)

No. 5. Histological typing of odontogenic tumours, jaw cysts, and allied lesions (1971)

No. 6. Histological typing of bone tumours (1972)

No. 7. Histological typing of salivary gland tumours (1972)

No. 8. Nomenclature of cytology of the female genital tract (1973)

No. 9. Histological typing of ovarian tumours (1973)

No. 10. Histological typing of urinary bladder tumours (1973)

No. 11. Histological typing of thyroid tumours (1974)

No. 12. Histological typing of skin tumours (1974)

No. 13. Histological typing of female genital tract tumours (1975)

No. 14. Histological and cytological typing of neoplastic diseases of haematopoietic and lymphoid tissues (1976)

No. 15. Histological typing of intestinal tumours (1976)

No. 16. Histological typing of testis tumours (1977)

CONTENTS

General preface to the series 9

Introduction 11

Cytology of non-gynaecological sites 13
    Pulmonary cytology 13
    Urinary tract cytology 14
    Body fluids cytology 15
    Gastro-oesophageal cytology 16
    Fine needle aspiration cytology 18

Explanatory notes 19
    Pulmonary cytology 19
    Urinary tract cytology 30
    Body fluids cytology 37
    Gastro-oesophageal cytology 39
    Fine needle aspiration cytology 44

Annex 1. Urine and prostatic secretions 53

Annex 2. Gastro-oesophageal specimens 55

Annex 3. Technique of aspiration biopsy of palpable lesions 57

Index 59

Colour microphotographs

GENERAL PREFACE TO THE SERIES

Among the prerequisites for comparative studies of cancer are internationa agreement on histological criteria for the classification of cancer types and a standardized nomenclature. At present, pathologists use different terms for the same pathological entity, and furthermore the same term is sometimes applied to lesions of different types. An internationally agreed classification of tumours, acceptable alike to physicians, surgeons, radiologists, pathologists and statisticians, would enable cancer workers in all parts of the world to compare their findings and would facilitate collaboration among them.

In a report published in 1952,[1] a subcommittee of the WHO Expert Committee on Health Statistics discussed the general principles that should govern the statistical classification of tumours and agreed that, to ensure the necessary flexibility and ease in coding, three separate classifications were needed according to (1) anatomical site, (2) histological type, and (3) degree of malignancy. A classification according to anatomical site is available in the International Classification of Diseases.[2]

The question of establishing a universally accepted classification by histological type has received much attention during the last 20 years and a particularly valuable Atlas of Tumor Pathology—*already numbering more than 40 volumes—is being published in the USA by the Armed Forces Institute of Pathology under the auspices of the National Research Council. An* Illustrated Tumour Nomenclature *in English, French, German, Latin, Russian, and Spanish has also been published by the International Union Against Cancer (UICC).*

In 1956 the WHO Executive Board passed a resolution[3] requesting the Director-General to explore the possibility that WHO might organize centres in various parts of the world and arrange for the collection of human tissues and their histological classification. The main purpose of such centres would be to develop histological definitions of cancer types and to facilitate the wide adoption of a uniform nomenclature. This resolution was endorsed by the Tenth World Health Assembly in May 1957[4] and the following month a Study Group on Histological Classification of Cancer Types met in Oslo to advise

---

[1] *Wld Hlth Org. techn. Rep. Ser.*, 1952, No. 53, p. 45.

[2] World Health Organization (1977) *Manual of the International Statistical Classification of Diseases Injuries, and Causes of Death*, 1975 revision, Geneva.

[3] *Off. Rec. Wld Hlth Org.*, 1956, 68: 14 (Resolution EB17.R.40).

[4] *Off. Rec. Wld Hlth Org.*, 1957, 79: 467 (Resolution WHA10.18).

WHO on its implementation. *The Group recommended criteria for selecting tumour sites for study and suggested a procedure for the drafting of histological classifications and testing their validity. Briefly, the procedure is as follows:*

For each tumour site, a tentative histopathological typing and classification is drawn up by a group of experts, consisting of up to ten pathologists working in the field in question. An international reference centre and a number of collaborating laboratories are then designated by WHO to evaluate the proposed classification. These laboratories exchange histological preparations, accompanied by clinical information. The histological typing is then made in accordance with the proposed classification. Subsequently, one or more technical meetings are called by WHO to facilitate an exchange of opinions and the classification is amended to take account of criticisms.

In addition to preparing the publication and the photomicrographs for it, the reference centre produces up to 100 sets of microscope slides showing the major histological types for distribution to national societies of pathology.

Since 1958, WHO has established 23 centres covering tumours of the lung; breast; soft tissues; oropharynx; bone; ovaries; salivary glands; thyroid; skin; male urogenital tract; jaws; female genital tract; stomach and oesophagus; intestines; central nervous system; liver, biliary tract and pancreas; upper respiratory tract; eye; and endocrine glands; as well as oral precancerous conditions; the leukaemias and lymphomas; comparative oncology; and exfoliative cytology. This work has involved more than 300 pathologists from over 50 countries. A number of these centres have completed their work, and most of their classifications have already been published (see page 6).

The World Health Organization is indebted to the many pathologists who have participated and are participating in this large undertaking. The pioneer work of many other international and national organizations in the field of histological classification of tumours has greatly facilitated the task undertaken by WHO. Particular gratitude is expressed to the National Cancer Institute, USA, which, through the National Research Council and the USA National Committee for the International Council of Societies of Pathology, is providing financial support to accelerate this work. Finally, WHO wishes to record its appreciation of the valuable help it has received from the International Council of Societies of Pathology (ICSP) in proposing participants and in undertaking to distribute copies of the classifications, with corresponding sets of microscope slides, to national societies of pathology all over the world.

INTRODUCTION

During recent years diagnostic cytology has been the most rapidly developing area in the field of pathology. General acceptance of cytology as an important method in the management of patients with cancer and allied diseases continues to grow. Although early detection of asymptomatic cancer in sites other than the female genital tract has been so far disappointing, screening by cytology can play a role in selected high-risk patients. Neoplasms of the lung, oral cavity, urinary bladder and gastrointestinal tract may be detected cytologically when little, if any, clinical evidence exists.

The major role of cytology in non-gynaecological sites is to establish proof of the existence of cancer where there is suspicion. Furthermore, the state of the art is rapidly reaching the point where specific diagnoses of tumour types can be made with a high degree of accuracy. Negative cytologic results do not exclude the presence of cancer.

In the selection of material for this publication the WHO Collaborating Centre has included examples of a variety of non-neoplastic diseases that enter into the differential diagnosis and, at times, present real diagnostic problems. We recognize that examples of all neoplasms of the various sites cannot be illustrated. This is especially true for aspiration cytology, which has developed to the point that nearly all sites are accessible to the fine needle.

The limitations of the cytologic method should not be ignored. Although the sampling techniques are relatively simple, they are not necessarily always performed proficiently, and less than an optimum sample may be submitted. The processing techniques are equally important and should not be delegated to the inexperienced or handled in a casual manner. Prompt processing, with good fixation and staining, are essential. Techniques vary but we have chosen, for the most part, to illustrate cells fixed promptly in alcohol or alcohol-ether and stained by the Papanicolaou technique.

The members of the Collaborating Centre are in agreement that the reporting of specimen interpretation by the cytopathologist should be in diagnostic terms and never by numerals. When possible the tumour type should be indicated as well. Sometimes the cytologic evidence may only indicate the presence of a malignant neoplasm without determining the cell type. Even such a diagnosis is helpful in patient management.

Since the implications of therapy of neoplastic disease may be drastic, one must guard against over-diagnosis when the evidence is not conclusive. In any debatable case histologic confirmation should be sought. Inadequate specimens must be so identified and a negative report should never be construed as indicating the absence of neoplasm.

The examination of respiratory, gastric and urinary tract specimens can be time-consuming. Well-trained cytotechnologists are invaluable in pre-screening the slides and marking the cells in question.

CYTOLOGY OF NON-GYNAECOLOGICAL SITES

Pulmonary Cytology

A. BENIGN

   Squamous metaplasia

B. ATYPICAL SQUAMOUS METAPLASIA (DYSPLASIA) AND CARCINOMA IN SITU

   1. Atypical squamous metaplasia
      (a) mild
      (b) moderate
      (c) severe
   2. Carcinoma in situ

C. MALIGNANT

   1. Squamous cell carcinoma
   2. Small cell carcinoma
   3. Adenocarcinoma
      (a) bronchogenic
      (b) bronchiolo-alveolar
   4. Large cell carcinoma

D. SECONDARY TUMOURS

E. UNCLASSIFIED TUMOURS

F. MISCELLANEOUS FINDINGS

Urinary Tract Cytology

I. BLADDER

A. BENIGN

   1. Metaplastic squamous cells
   2. Malakoplakia
   3. Inflammation

B. CARCINOMA IN SITU

C. MALIGNANT

   1. Epithelial
      (*a*) Papillary urothelial carcinoma
      (*b*) Non-papillary urothelial carcinoma
      (*c*) Squamous cell carcinoma
         (*i*) large cell non-keratinizing
         (*ii*) keratinizing
      (*d*) Adenocarcinoma
      (*e*) Undifferentiated carcinoma

   2. Non-epithelial
      Sarcomas [1]

D. SECONDARY TUMOURS [1]

E. UNCLASSIFIED TUMOURS

II. PROSTATE

Adenocarcinoma

III. RENAL PARENCHYMA

Carcinoma

---

[1] Also found in kidney and ureter.

Body Fluids Cytology

A. BENIGN
   Mesothelial cells

B. MALIGNANT
   Mesothelioma

C. SECONDARY TUMOURS

D. UNCLASSIFIED TUMOURS

Gastro-oesophageal Cytology [1]

I. OESOPHAGUS

A. BENIGN

    1. Epithelial

        Squamous cell papilloma

    2. Non-epithelial

B. MALIGNANT

    1. Epithelial

        (a) Squamous cell carcinoma

        (b) Adenocarcinoma

        (c) Undifferentiated carcinoma

        (d) Others

    2. Non-epithelial

C. MISCELLANEOUS TUMOURS

    1. Carcinosarcoma

    2. Malignant melanoma

    3. Others

D. SECONDARY TUMOURS

E. UNCLASSIFIED TUMOURS

II. STOMACH

A. BENIGN

    1. Epithelial

        Adenoma

    2. Non-epithelial

---

[1] These classifications generally adhere to those of the WHO Collaborating Centre for the histological classification; modifications are introduced where appearances differ specifically relating to cytology.

B. MALIGNANT

1. Epithelial

    (*a*) Adenocarcinoma
      (*i*) well differentiated
      (*ii*) moderately differentiated
      (*iii*) poorly differentiated
      (*iv*) signet-ring cell carcinoma

    (*b*) Adenosquamous carcinoma

    (*c*) Squamous cell carcinoma

    (*d*) Undifferentiated carcinoma

2. Non-epithelial

C. MISCELLANEOUS TUMOURS

1. Carcinoid tumours

2. Haematopoietic and lymphoid neoplasms

    (*a*) Lymphosarcoma

    (*b*) Reticulosarcoma

    (*c*) Hodgkin's disease

    (*d*) Others

D. SECONDARY TUMOURS

E. UNCLASSIFIED TUMOURS

Fine Needle Aspiration Cytology

For classifications, the relevant texts in the World Health Organization series *International Histological Classification of Tumours* should be consulted, as follows:

Thyroid	No. 11. *Histological typing of thyroid tumours*, by Chr. Hedinger & L. H. Sobin (1974)
Breast	No. 2. *Histological typing of breast tumours*, by R. W. Scarff & H. Torloni (1968)
Prostate	Text in production
Haematopoietic and lymphoid tissues	No. 14. *Histological and cytological typing of neoplastic diseases of haematopoietic and lymphoid tissues*, by G. Mathé, H. Rappaport, G. T. O'Conor & H. Torloni (1976)

EXPLANATORY NOTES

Pulmonary Cytology

Cytologic diagnosis of lung cancer is based largely on the detection of exfoliated carcinoma cells in sputum or bronchial secretions. The procedure can be employed as either a diagnostic or a screening technique. As currently used, pulmonary cytologic studies constitute an important step in the diagnostic work-up of patients with pulmonary symptoms or with an X-ray abnormality of the chest. Unfortunately, such patients are generally diagnosed at a relatively advanced stage of the disease and consequently the results of treatment of bronchogenic carcinoma up to the present time have been disappointing (overall cure rate 5%). However, there is evidence that the earlier the disease can be localized and treated the better the prognosis. Optimum improvement might be expected if bronchial carcinoma could be detected in its asymptomatic, X-ray-negative stage. Cytologic screening of high-risk asymptomatic subjects would appear to offer hope of earlier diagnosis and the feasibility of such screening as a protection against lung cancer, particularly in the older smoking population, is currently under study in a number of centres.

Examination of sputum is the most widely used procedure in pulmonary cytology. Collection of adequate specimens is of cardinal importance and to a large extent success or failure of the method depends on this procedure. The most satisfactory specimens are obtained when the patient is carefully instructed to produce a morning deep cough sample. Examination of multiple morning specimens is recommended and it is generally considered that a series of 3–5 specimens constitutes an adequate cytologic study. In general, the more specimens examined the higher the percentage of positive results.

Bronchial washings or secretions are obtained during bronchoscope examinations and these are processed and examined as for sputum specimens. Although either type of specimen may be positive in a given case, it is generally felt that bronchial washings or aspirates are less productive of positive results than sputum.

In difficult problems such as the cytologic diagnosis of small, peripherally placed lung carcinomas or the localization of occult, X-ray-negative in-situ or early invasive bronchogenic carcinomas, special techniques may be used to obtain diagnostic cytologic specimens. These include bronchial brushing using the fibreoptic bronchoscope under X-ray television guidance, as well as the use of transthoracic fine needle aspiration of peripheral pulmonary nodules.

With a series of 3–5 specimens of sputum, 70–80% of bronchogenic carcinomas can be diagnosed cytologically. When all of the above techniques are used, a positive cytologic diagnosis is likely in nearly all bronchogenic cancers. For excellence in performance, the cytopathologist needs considerable special training and experience. False-positive reports under these conditions should be extremely rare. The cytopathologist must be well acquainted with benign cellular alterations which may mimic cancer, as well as with contaminants such as vegetable cells, which are also sometimes a source of error.

THE CELLULAR CONSTITUENTS OF PULMONARY CYTOLOGY SPECIMENS

The *benign cells* encountered in pulmonary specimens include a mixture of epithelial and inflammatory cells in various stages of preservation. Epithelial cells include squamous cells derived mainly from the oral mucosa, high columnar cells from the lower respiratory tract, small columnar cells from the upper respiratory tract and occasionally parabasal and reserve cells. Variations of these epithelial cells may be observed, including columnar cells with a modified shape and multinucleation, and goblet cells with or without intracellular mucus. In addition, metaplastic squamous cells exhibiting varying degrees of atypia may be present. Common inflammatory cells include polymorphonuclear leucocytes, eosinophils, occasional lymphocytes and plasma cells, and various forms of histiocytes and macrophages.

The *malignant cells* encountered in pulmonary cytology specimens include carcinoma cells derived from primary or metastatic lung cancer and occasionally cells from such rare primary or secondary tumours as malignant lymphoma and sarcoma.

TUMOUR TYPING

In addition to the diagnosis of carcinoma in pulmonary cytology specimens an attempt should be made to classify positive cases according to histologic tumour type. Cytoplasmic alterations in tumour cells are often helpful in this procedure, as well as such factors as size and arrangement of malignant cells. In general, the cytologic classification herein provided follows the WHO histopathologic classification, although the cytologic subtyping is less detailed in scope. Under optimal conditions, an accurate subclassification can be made of cancer cells into squamous cell carcinoma, small cell carcinoma, adenocarcinoma (including bronchiolo-alveolar cell carcinoma), and large cell carcinoma. A number of poorly differentiated carcinomas are difficult to classify. It is generally considered that an accurate classification as to type is possible in at least 80% of positive cases. Special tumours of bronchial gland origin, such as carcinoid, rarely desquamate. Other primary tumours such as lymphoma and carcinosarcoma occur too rarely in pulmonary cytology material to be included in a formal classification.

Secondary tumours in the lung can also be diagnosed cytologically.

BENIGN CELLS IN PULMONARY SPECIMENS [1]

SPUTUM

Sputum is expectorated material containing cells from the lower respiratory tract. Sputum is characterized by the presence of alveolar histiocytes. Respiratory epithelial columnar cells are infrequent in a spontaneously raised sputum sample. Sputum obtained by artificial stimulation of the cough reflex may contain isolated or small groups of columnar cells. Sputum obtained after bronchoscopy also often exhibits an increase of columnar cells. The presence of ciliated columnar cells is not an absolute indication of material originating from the lower respiratory tract, as it may not be possible to distinguish whether such cells originate from the upper or lower respiratory tract.

SALIVA

Saliva represents watery salivary gland secretion and contains mainly oral squamous cells. Examination of saliva has no value in the diagnosis of pulmonary lesions.

BRONCHIAL ASPIRATES AND WASHINGS

Bronchial aspirates and washings are obtained by bronchoscopy. The cellular composition is different from that of sputum samples. The predominant cell type in the bronchial aspirate (or washing) is ciliated columnar cells from the surface layer of the bronchial epithelium. These cells occur singly or in loose or tight clusters. Oral squamous cells and alveolar macrophages are usually few in number or may be absent. Goblet cells are frequently seen in these specimens, while inflammatory cells occur less frequently than in sputum specimens. The number of cells is, in general, more limited in bronchial aspirates and washings than in sputum.

---

[1] Papanicolaou stain.

SQUAMOUS EPITHELIAL CELLS (upper respiratory origin)

Three cell types from the upper respiratory passage occur in pulmonary specimens:

1. *Superficial cells* are large, thin and polygonal with transparent, usually eosinophilic, cytoplasm. The nucleus is pyknotic.

2. *Intermediate cells* are from the intermediate layers of the squamous epithelium. They are similar in size to the superficial cells. There is a polygonal outline and an eosinophilic or cyanophilic cytoplasm. The nucleus is vesicular. Intermediate eosinophilic cells are the predominant cell type in sputum specimens [1].

3. *Parabasal cells* arise from the parabasal layer of the epithelium. The cytoplasm is dense, has a well demarcated outline and stains cyanophilic. The nuclei are similar to those of the intermediate cells. Parabasal cells are extraordinarily infrequent in exfoliated material.

RESPIRATORY EPITHELIAL CELLS

There are three types of respiratory epithelial cells:

1. *High columnar ciliated cells*, which vary in length. They exhibit one broader end with a flat and thickened cell membrane (terminal plate) which carries the cilia. The cilia as well as the terminal plate often stain pink. The opposite end of the cell is tapered. The cytoplasm is homogenous and faintly cyanophilic. The nucleus is located in the middle of the cells or somewhat closer to the tapered end. It is oval or round, vesicular with one or two fine clumps of chromatin, and contains a nucleolus which occasionally is distinct or even prominent. Multinucleation is not uncommon. The columnar cells occur singly or in clusters or as small epithelial fragments of the superficial layer of the mucosa. A normal spontaneously raised sputum contains few or no such cells, while a bronchial aspiration specimen or washing usually is dominated by columnar cells.

Multinucleation in columnar cells is not unusual. The identification of a multinucleated columnar cell may, however, sometimes be difficult, especially when the columnar shape is modified and the nuclei are darkly stained. Cilia are often not visible. In such cases there are some identification signs which are of utmost importance for determining the cell type. A suggestion of a columnar, cubic or triangular shape, a terminal plate, or remnants of cilia at one cellular border are all strong indications of origin in benign columnar epithelium. The existence of one dome-shaped side of such a cell is also an important identification sign; a delicate terminal plate often assumes such a shape.

---

[1] Though not usually predominant in actual number of cells present, the presence of alveolar macrophages is the important indicator of a satisfactory sputum specimen.

Nuclear abnormalities, mainly enlargement and hyperchromasia, but rarely a coarse chromatic pattern, may be encountered in columnar cells. If there is a modification of the columnar shape as well, differential diagnostic problems may arise. In such instances it is important to look for the various identification signs described above.

2. *Goblet cells* are also exfoliated from the epithelial surface. They are similar in size to columnar cells. The goblet cells carry no cilia. The cytoplasm is faintly cyanophilic and often vacuolated and distended by mucus. In this state the nucleus is often pushed to the distal, tapered end of the cell and assumes a triangular shape. When the cytoplasm contains little or no mucus it may exhibit the appearance of a rumpled paper bag rather than a goblet.

3. *Cells of the germinative layer of the epithelium*, i.e., basal or reserve cells, are extremely rare in exfoliated material. These are small, round, oval or polygonal cells. The cytoplasm is scanty and cyanophilic. The nucleus is round or oval and vesicular.

METAPLASTIC SQUAMOUS CELLS

Metaplastic squamous cells are larger than parabasal cells but smaller than oral squamous cells and have a cuboidal, polygonal or round outline. The cytoplasm is usually eosinophilic, or more rarely cyanophilic. It exhibits some transparency but is not so thin and delicate as that of oral squamous cells, and not so dense as that of keratinized cells. The nucleus of metaplastic squamous cells is usually round and regular, with a fine chromatin network. The nuclear size is the same as, or somewhat larger than, that of oral squamous cells. Metaplastic squamous cells are easily recognizable when occurring in clusters, particularly when one side forms a flat surface or when some columnar cells are attached to the end of the cluster. Such features produce a resemblance to the surface lining of the bronchi. Metaplastic squamous cell clusters exhibit a marked degree of cellular adhesiveness.

Metaplastic squamous cells may exhibit varying degrees of nuclear atypia.[1] In mild and moderate atypical squamous metaplasia these changes include nuclear enlargement and hyperchromasia as well as some variation in nuclear size and shape. In the more severe atypias there is a more pronounced hyperchromasia of the nuclei, the chromatin may be more granular, and nuclear size and shape more variable. In the most extreme examples the nuclear changes approach those seen in carcinoma. In reporting cytologic preparations containing metaplastic squamous cells, the following classification is suggested: (1) metaplastic squamous cells with no atypia, (2) metaplastic squamous cells with mild atypia, (3) metaplastic squamous cells with moderate atypia, and (4) metaplastic squamous cells with severe atypia.

---

[1] The term dysplasia may be used as an alternative term to atypical squamous metaplasia.

In all of the above gradations of identifiable squamous metaplasia, the cytoplasm is predominantly eosinophilic. Orangeophilia, indicating keratin production, is not usually seen in metaplastic squamous cells.

There is evidence of a correlation between the presence of metaplastic squamous cells in sputum and the corresponding squamous metaplasia of bronchial mucosa as established by tissue sections. Metaplastic squamous cells with severe atypia are presumably derived from an epithelial surface which approaches the histologic picture of epidermoid carcinoma in situ.

PULMONARY MACROPHAGES

Pulmonary macrophages (alveolar macrophages, dust cells) are large histiocytes coming from the alveoli of the deep respiratory passages. The occurrence of such cells indicates that a specimen is an adequate sputum sample, i.e., that material from the deep parts of the respiratory tract has been obtained. The pulmonary macrophages are round or oval and usually larger than ordinary histiocytes. They measure 10–30 μm or more in diameter.

There is a distinct cytoplasmic border. The cytoplasm is cyanophilic, eosinophilic or amphophilic. It often contains a few or many grey, brown, or black particles. The cytoplasm may be finely vacuolated, especially in the absence of such particles, and may be so full of foreign material that the macrophages appear as brown or black ovoids without visible nuclei.

The nuclei of alveolar macrophages are usually of the same size as those of oral squamous cells. They are vesicular, with a fine, uniform chromatin pattern. The nuclei are mostly round, but may exhibit a discus shape. The alveolar macrophages are usually mononuclear, but binucleation is common and multinucleation is occasionally seen.

INFLAMMATORY CELLS

Inflammatory cells occur in the background of respiratory cytologic specimens, particularly sputum. They vary greatly in number and state of preservation. Usually they provide no specific diagnostic information. However, a purulent specimen may indicate severe pulmonary inflammation; numerous eosinophils may relate to asthmatic bronchitis, and epithelioid cells are often associated with specific granulomatous disease, particularly tuberculosis.

1. *Small histiocytes* of the same type as in other cytologic material may occur in sputum and bronchial secretions. These cells are usually round, but may be slightly irregular. The nuclei are often eccentric, with a round or ovoid shape, and stain light blue or grey. In a cluster of small histiocytes there is usually one or more cells with typical bean- or kidney-shaped nuclei,

which are an important identification sign. The cytoplasm is cyanophilic and may be finely vacuolated and contain ingested particles.

2. *Giant multinucleated histiocytes* of foreign body type are 2–10 times the size of ordinary small histiocytes. The nuclei are round, with a fine chromatin network; they are evenly distributed but may occasionally be concentrated centrally. There may be up to 100 nuclei, though 20–40 is more common. The cytoplasm is abundant and stains cyanophilic or amphophilic. The cell border is well defined.

3. *Epithelioid cells* are mononuclear phagocytes found around necrosis in tuberculous or other specific granulomata. These cells show a faint resemblance to epithelial cells in histologic sections. In the sections the cells are oval or elongated with abundant, pink cytoplasm and poorly defined margins.

In cytologic specimens it is usually the elongated forms which are seen. These cells show great similarity to ordinary histiocytes and appear in clusters together with other types of histiocytic cells. Transitional forms of elongated and rounded types are frequent. They have poorly defined, finely vacuolated cytoplasm which stains amphophilic or cyanophilic. The nuclei are elongated and often folded. They may have indentations, tapered ends, and sometimes a modified kidney- or bean-shape; a pear-shape is also often seen. Langhans giant cells differ from the foreign body type mainly in the peripheral location of the nuclei, which often overlap and are slightly elongated.

MALIGNANT CELLS IN PULMONARY SPECIMENS

SQUAMOUS CELL CARCINOMAS

These tumours are characterized by keratin production or intercellular bridges. A variation in the degree of maturation is indicated by the amount of keratin production and the number of cells undergoing keratinization. Thus, the carcinomas may be described as highly, moderately, or poorly differentiated. Squamous cell carcinomas may occasionally be detected in an early, predominantly in situ stage, though more commonly in the infiltrative stage.

1. *In situ squamous cell carcinoma*

In situ, or very early, bronchogenic carcinoma may be diagnosed by cytologic methods at an X-ray-negative stage and the tumour subsequently localized by bronchoscopic examination. Gross abnormalities may be inconspicuous. Histologically, in the in situ stage the carcinomatous changes are limited to the surface epithelium and ducts of mucus glands. In addition to the intraepithelial changes, early cases show incipient focal microscopic infiltration of the submucosa (microinvasion).

2. *Invasive squamous cell carcinoma*

In cytologic specimens, the most distinctive finding in highly differentiated cases is the presence of tumour cells of varying size with abundant orangeophilic cytoplasm indicative of keratin production. The tumour cells may occur singly, in small clumps or loose clusters. Much variation in size and shape is characteristic of this tumour type and bizarre forms or elongated spindle-shaped cells are seen. Marked nuclear hyperchromasia with coarse chromatin is uniformly present and some nuclei may appear pyknotic, as the result of either maturation or degenerative change. Occasionally one finds an " epithelial pearl ".

Cytologic specimens from squamous cell carcinoma of a slight or only moderate degree of differentiation exhibit a lesser tendency to keratin production, hence fewer orangeophilic cells are seen and a relatively higher percentage of cells show amphophilic or cyanophilic cytoplasm. The tumour cells are in general somewhat less bizarre, with less variation in size and shape. The nuclear chromatin is coarsely granular and prominent nucleoli may be present.

Cytologic specimens from in situ bronchogenic carcinoma (or in situ carcinoma with associated early invasion) are frequently sufficiently distinctive. These specimens are characterized by a variable, often large, number of well-differentiated abnormal cells of squamous type. Within this spectrum of abnormal squamous cells are: (1) carcinoma cells which fulfil the requirements for squamous cell carcinoma, possibly less bizarre than those of classical invasive type; (2) large cells which are polygonal, round or irregular with abundant, usually orangeophilic or eosinophilic cytoplasm and enlarged, slightly hyperchromatic nuclei; and (3) small abnormal squamous cells, usually round to oval, with evidence of cell keratinization. In the latter the nuclei are round or slightly irregular, with varying degrees of hyperchromasia and coarseness of chromatin.

The above abnormal cells tend to occur singly, but clusters with a fair degree of cohesiveness may be present. The usual background seen in invasive carcinoma, consisting of masses of cellular débris, inflammatory exudate and altered or fresh blood, is lacking in the majority of cases of early and in situ carcinoma.

SMALL CELL CARCINOMA

These tumours are composed of small cells without differentiation. There is a particularly good correlation between the histopathologic section and the typical cytologic findings in sputum. In cytologic specimens the most characteristic and diagnostic feature is the small size of the tumour cell, which consists largely of a nucleus with an inconspicuous rim of cyanophilic cytoplasm. The cells are arranged in loose clusters with a scattering of single cells. Nuclear hyperchromasia is striking, with a " sievelike " or coarsely granular chromatin pattern. Nucleoli are typically absent. A moderate degree of variation in size and shape of nuclei is usually present and nuclear moulding may be noted. Some variation in the overall size of cells is recognized and somewhat smaller than average or larger cell variants are seen.

ADENOCARCINOMAS

The tumours are characterized by the formation of gland-like spaces, with or without papillary structure, and/or the presence of mucus in the tumour cells. Further subdivision is possible according to the degree of differentiation of the carcinoma and the site of origin and mode of growth within the lung parenchyma.

1. *Bronchogenic adenocarcinoma*

In cytologic specimens, tumour cells from bronchogenic adenocarcinoma vary according to the degree of differentiation, and tend to occur in clumps, single cells being found less frequently. These can be diagnosed as adenocarcinoma cells in a large proportion of cases, especially when the tumour shows either a moderate to marked degree of glandular differentiation or a papillary architecture. Characteristically, the tumour cells are large, as compared to bronchial epithelial cells, with a relatively large nucleus and altered nuclear/cytoplasmic ratio. Nuclear changes include moderate hyperchromasia with coarse chromatin and large prominent nucleoli. The cytoplasm is moderate in amount and usually somewhat cyanophilic. Intracytoplasmic mucus may take the form of small or large vacuoles or, infrequently, have a bubbly or foamy appearance. Many qualitative variations of these cytologic changes are seen, depending on the degree of differentiation of the tumour. Cytologic specimens from poorly differentiated adenocarcinoma cannot be distinguished readily from large cell carcinoma.

2. *Bronchiolo-alveolar adenocarcinoma*

These highly differentiated, usually papillary, adenocarcinomas occur in the periphery of the lung and spread throughout the lung parenchyma, using the alveolar walls as framework. A large alveolar surface may be involved by papillary neoplasm, providing ample opportunity for desquamation of tumour cells. Cytologic specimens from such carcinomas frequently show numerous clumps of well-differentiated tumour cells, rather uniform in size and shape and often arranged in acinar or somewhat elongated papillary fashion. The diagnosis of bronchiolo-alveolar carcinoma depends on this finding together with slight to moderate cellular changes. These include slight increase in nuclear size (in comparison with the background of bronchial epithelial cells), and moderate nuclear hyperchromasia, with or without visible nucleoli.

LARGE CELL CARCINOMA

These tumours are composed of large cells with no evidence of differentiation. Some variation in structure is seen within this subgroup, in particular as regards size of cells and amount of cytoplasm. Clear cell and giant cell variants occur.

The cytologic findings in sputum reflect the structure of the tumour. The cells occur singly or in small groups and are generally large and anaplastic with striking nuclear changes including large nuclear size, irregularities of nuclear membrane, and coarse irregular chromatin masses. Enlarged, round or irregularly shaped nucleoli may be noted. The amount of cytoplasm varies from slight to moderate, is cyanophilic or amphophilic and shows no evidence of differentiation. As is true of histologic preparations, a diagnosis of large cell carcinoma on a cytologic basis is largely one of exclusion. Cells exfoliated from poorly differentiated adenocarcinoma or squamous cell carcinomas are thus likely to be confused with those from large cell carcinoma.

MISCELLANEOUS FINDINGS

1. *Viral changes*

The changes associated with herpes virus infection involve squamous and columnar cells. There is cellular and nuclear enlargement, with multinucleation as a common feature, often resulting in formation of characteristic giant cells. Nuclear crowding results in moulding of adjacent nuclei. There is an opaque homogenization of chromatin, resulting in a ground-glass appearance of nuclei, often with a characteristic margination of chromatin material. The centrally located large intranuclear inclusion bodies often found in gynaecological specimens are rarely visible in pulmonary specimens.

2. *Contaminants*

Vegetable cells may occur as contaminants of sputum and on occasion may mimic carcinoma cells. They are generally uniform in size, may be arranged in palisade form and show prominent cellulose membranes with acidophilic refractile granules within the cytoplasm. The nuclei usually stain darkly and, in contrast to tumour cells, are without internal structure.

3. *Curschman's spirals*

Curschman's spirals are coiled structures frequently seen in sputum, particularly in patients with obstructive lung disease. The spiral is composed of inspissated mucus and consists of a dark central axis surrounded by a translucent mucoid envelope.

4. *Ferruginous (asbestos) bodies*

These structures in sputum are commonly associated with exposure to asbestos. The ferruginous body (10–80 μm in length) is thick and brownish due to haemosiderin deposition and has transverse corrugations and clubbed ends.

Urinary Tract Cytology

Urinary diagnostic cytology is of importance for the detection of neoplasia in three areas: urinary tract, prostate and renal parenchyma. Of these three the urinary tract is by far the most frequently affected.

For the understanding of carcinogenesis of the urothelium, it is important to realize that the presence of a neoplasm is often associated with widespread epithelial changes throughout the urothelium which may determine the subsequent course of the disease. The urothelial neoplasms (transitional cell tumours) are most frequently encountered in the urinary bladder; the renal pelvis is second in order of frequency, while ureter and male urethra are the least commonly affected sites. It is possible to detect urothelial neoplasia in its earliest stage, i.e., in the stage of carcinoma in situ, when there is anaplasia of the epithelium without the formation of papillary structures and without infiltration.

Prostatic tumour cells can be found in voided urine. Certain workers add to the diagnostic procedure the examination of material obtained by prostatic massage and examination of the first subsequent urinary sample. Prostatic massage in patients not otherwise suspected of tumour may lead to an early diagnosis. The fine needle aspiration mentioned elsewhere is a further technique for confirming the presence of neoplasia.

The renal parenchymatous tumours may exfoliate cells through the renal tubules into the urinary tract, thus allowing a diagnosis to be made at a relatively early stage of renal carcinoma. It is, however, more frequent to find cells of renal carcinoma in the urine when the tumour is advanced and has involved the renal pelvis.

Urothelial carcinogenesis is of particular significance for certain high-risk groups of the population, namely those who work in certain chemical industries, the heavy smoker and the patient with schistosomiasis. Aniline dyes, benzidine, β-naphthylamine and other chemicals, when absorbed, may lead, after an interval of many years, to urothelial changes which may be present many years before the tumour manifests itself. In some of these workers the first lesion may be a carcinoma in situ, which can be detected by routine screening of the exposed workers. Patients with schistosomiasis also frequently develop urothelial neoplasia, usually of squamous cell type. A direct etiological relationship between schistosomiasis and urothelial neoplasia has, however, not been established with certainty.

It is therefore suggested that the following signs and clinical histories are indications for urine cytodiagnosis:

(1) haematuria and microhaematuria;

(2) persistent urological symptoms (chronic infection);

(3) high-risk groups (industrial workers in chemical and other industries);

(4) schistosomiasis; and

(5) follow-up after treatment for urinary tract tumours.

It must be appreciated that the urothelium may undergo changes as the result of the systematic administration of drugs. The use of cytotoxic drugs may lead to epithelial alterations almost indistinguishable from carcinoma.

It is therefore imperative that all specimens of urine for cytodiagnosis should be accompanied by an adequate clinical history which includes data on treatment with drugs and radiation.

BENIGN CELLS IN URINE SPECIMENS

NORMAL CELLS

Normal bladder urine contains very few epithelial cells and no bacteria, erythrocytes or inflammatory cells. The urothelial cells have pale cyanophilic cytoplasm, distinct cytoplasmic borders and round nuclei with fine granular chromatin. The nuclei are central or eccentric and the nuclear/cytoplasmic ratio is approximately 1 : 3. The cells are comparable in size with the parabasal cells of squamous epithelium and may have long extensions and irregularities in shape. A few eosinophilic or orangeophilic squamous cells from the urethral meatus are seen in voided specimens from males. Non-catheterized specimens from females always contain vaginal squamous and inflammatory cells and organisms. For this reason, definitive cytologic study in women with urinary diagnostic problems is best carried out on catheterized specimens or specimens obtained by transpubic puncture.

The urothelium in the base of the bladder is influenced to some extent by oestrogenic substances to differentiate along squamous lines. This occurs to a variable degree in women during the menstrual cycle or as the result of oestrogenic therapy in men or women. If it is desirable to study this change for any reason, accurate evaluation requires catheterized bladder urine specimens from both sexes.

Renal tubular epithelial cells are seen singly or in groups of 4–5 cells in renal parenchymal disease or circulatory disturbances resulting in loss of nephrons. They are round or cuboidal, with indistinct cytoplasmic borders and rather dense faintly granular, grey or reddish-brown cytoplasm and round nuclei often showing degenerative changes. Tubular casts of various types often accompany these.

Special caution is necessary in interpreting cellular changes if the patient has been subjected to radiation therapy to the pelvis or abdomen or is receiving cytotoxic drugs. Unusually large nuclei and variation in cell size occur in these circumstances. Even the common analgesic drugs have been shown to cause excessive desquamation of renal tubular epithelial cells which are rather large, pleomorphic and sometimes binucleated.

Traumatic evulsion of normal urothelium by ureteral catheter produces many pleomorphic urothelial cells and groups of cells indistinguishable from those of low-grade papillary carcinoma. Such cells found in ureteral catheter specimens should be considered significant only if they have been observed in previous bladder urine specimens.

A further caution is warranted with respect to contamination of urinary or prostate specimens with seminal fluid. Large round or cuboidal dusky grey cells, at times containing brown pigment, from the epithelium lining the seminal vesicles are often found. Such cells have large hyperchromatic nuclei and may be mistaken for carcinoma cells. They do not have prominent nucleoli, however, and the presence of spermatozoa is the clue to their identity.

Occasionally, intact Brunn's nests with a central lumen containing secretion may be seen.

METAPLASTIC CELLS

Squamous metaplasia of the urothelium or prostatic duct epithelium is not uncommon and may be focal or diffuse, non-keratinizing or keratinizing. Small foci are sometimes related to the presence of kidney, bladder or prostatic stones or chronic prostatitis. The non-keratinized metaplastic cells develop a more angular, flattened shape with cytoplasmic staining qualities ranging from clear cyanophilia to pale eosinophilia. The nuclei are central and vesicular. In the extreme degree of keratinizing metaplasia, large, flat, angular cells are densely eosinophilic or orangeophilic. They may be anuclear or have central pyknotic nuclei.

In acute cystitis there may be scattered neutrophils and erythrocytes and small clumps of urothelial cells. Cytoplasmic vacuoles may be present. The nuclei are somewhat enlarged and contain sparse coarse chromatin granules and small eosinophilic nucleoli.

MALAKOPLAKIA

Malakoplakia of the bladder may be represented in the sediment by large macrophages containing Michaelis-Gutmann bodies.

INFLAMMATION

Chronic infection of the bladder, urethra, ureter or renal pelvis gives rise to urothelial cell changes deceptively similar to those of malignant cells. In specimens from patients with a history of chronic urinary infection and in specimens with a background containing many bacteria and inflammatory cells, the presence of urothelial cells showing nuclear enlargement, binucleation, eosinophilic nucleoli and sharply granular chromatin must be interpreted with caution, and the diagnosis of cancer made only on strong evidence of intranuclear abnormality.

Chronic cystitis cystica may reveal itself in the urinary sediment by the appearance of spherical nests of uniform urothelial cells, the outer members of which tend to encircle the group.

CARCINOMA IN SITU CELLS IN URINE SPECIMENS

The cytological picture of carcinoma in situ is variable in accordance with the different histological terminology, depending upon whether one includes or not under the heading " carcinoma in situ " the so-called unstable labile *maladie de la muqueuse* described by French authors. The members prefer the first alternative.

Thus one may find multiple single cells of a monotonous cell pattern with grossly enlarged and hyperchromatic pleomorphic nuclei which may simulate an invasive neoplasm, or one may see the cells described below for well-differentiated urothelial malignant tumours. The background is relatively clean as compared with invasive tumours.

In the absence of cystoscopic evidence of tumour and with the cytological picture described above, the diagnosis of carcinoma in situ may be suspected. Sometimes a " salmon-pink " area may give a hint of the presence of this lesion at cystoscopy. A definite diagnosis can only be made by multiple blind biopsies and histological diagnosis.

The carcinoma in situ resists radiation therapy. It may take months to years before a gross " tumour " develops. The latter may never occur and invasive growth may develop from carcinoma in situ without visible tumour formation.

MALIGNANT CELLS IN URINE SPECIMENS

UROTHELIAL (TRANSITIONAL CELL) TUMOURS

Neoplastic urothelial cells are mainly round, with large, often eccentric nuclei. The cytoplasm is cyanophilic or haemoglobin-stained and shows a well-defined border. Intranuclear abnormalities exceed those which may represent urothelial reaction to inflammatory or mechanical injury. Increasing loss of differentiation is indicated by cells with more anisocytosis, anisonucleosis and pleomorphism; monstrous and small deformed cells with lesser amounts of cytoplasm are seen. Invasion is indicated by a background of cellular debris, inflammatory exudate and altered and fresh blood. These indications of the extent of the tumour are particularly valid in bladder tumours, but less so when the tumour is located in the renal pelvis. In the latter case, the ratio of abnormal to normal cells is increased, although the total number of epithelial cells may be decreased. Anaplastic carcinomas may shed large numbers of small cells, fairly uniform in size and round or oval in shape, with a very thin rim of cyanophilic cytoplasm and hyperchromatic nuclei.

1. *Papillary urothelial carcinoma*

The cells are as described above but often display long cytoplasmic processes, particularly in well-differentiated varieties. Distinctive groups with cells arranged in a regular fashion around the periphery permit the diagnosis of a papillary tumour.

2. *Non-papillary urothelial carcinoma*

The cells are mainly single or in small, irregular groups. They vary in size but are less pleomorphic than those from papillary carcinoma.

3. *Squamous cell carcinoma*

Usually the cells are of the type consistent with large cell non-keratinizing carcinoma, less frequently with keratinizing carcinoma.

(*a*) *Large cell non-keratinizing carcinoma*. The cell study reveals syncytial clusters as well as isolated relatively large cyanophilic cells. The chromatin pattern is coarse and irregular and there are several dense chromocentres of varying size and shape. The nuclei contain prominent macronucleoli. Although cell size tends to be uniform, there may be moderate variation.

(*b*) *Keratinizing carcinoma*. There is a preponderance of relatively large abnormal cells with a high degree of pleomorphism characterized by caudate and elongated forms. The chromatin pattern is coarse and irregular and there are several dense chromocentres of varying size and shape. Macronucleoli are not prominent. Nuclear pyknosis is a distinct feature. The cytoplasm may be very abundant and is often eosinophilic or orangeophilic.

In cases of squamous cell carcinoma associated with schistosomiasis, keratinous débris may predominate and malignant cells typical of squamous cell carcinoma may be few in number.

4. *Adenocarcinoma*

This is rare as a primary tumour. The cytology is similar to that of adenocarcinoma in other sites.

5. *Undifferentiated carcinoma*

Most of the cells are as described in poorly differentiated tumours and may occur in disorganized clumps representing fragments of tumour. Very often among malignant urothelial and anaplastic cells there are large non-keratinizing squamous cells and/or glandular cancer cells, representing the variation in the cellular components of this tumour.

SARCOMA

Evidence of sarcomas of the urinary tract, usually rhabdomyosarcoma involving the bladder, ureter or kidney, is found in the urine only after ulceration of the overlying epithelium. Hence, the cells are seen in a background of blood and necrotic débris which may obscure many of their features. Highly abnormal, often pleomorphic nuclei with large, multiple nucleoli are found. The cells have little or no cytoplasm. When present, the cytoplasm is usually seen as fragile wisps or sometimes as a dense eosinophilic band or "strap". Striations are very rarely identified. The cytoplasmic borders are indistinct.

PROSTATIC ADENOCARCINOMA

The cells are found in three-dimensional clusters or singly. The cyanophilic cytoplasm is generally finely vacuolated but may be homogeneous. Large vacuoles are rarely found. The cells have distinct cytoplasmic borders. The nuclei, usually uniform in size, sometimes show pleomorphism. Some naked nuclei may be present. Nuclei are round or oval. The chromatin tends to be finely granular. The tumour background is often lacking.

Assessment of the degree of differentiation may be based on the fact that the less differentiated the tumour, the larger the cells, the nuclei, and the nucleoli.

RENAL PARENCHYMAL CARCINOMA

Cells exfoliate sparingly from these tumours in most cases but can be identified in a significant number of patients with the disease. The background contains variable amounts of old and fresh blood, but inflammatory cells are usually not numerous and bacteria are present only in small numbers throughout the specimen. Occasionally in advanced cases compact groups of cells are present. Usually renal carcinoma cells have variable amounts of opaque, indistinctly granular, cyanophilic, grey or brown cytoplasm without definite cytoplasmic borders. The nuclei are large and eccentric.

Sometimes the nuclei are extremely hyperchromatic but are often vesicular, with a few coarse chromatin granules and very prominent red nucleoli. They can be recognized even in an advanced stage of degeneration. In a few instances, those tumour cells in groups have the characteristics of small well-differentiated adenocarcinoma cells from other organs. These cells are difficult to differentiate from prostatic carcinoma cells, but the presence of a clear-cut cytoplasmic border in prostatic epithelial cells facilitates the diagnosis.

After the cells have been identified in bladder specimens, it may be necessary to catheterize the ureters separately in order to localize the source of the cells.

SECONDARY TUMOURS

These entities differ according to the part of the urinary tract which is involved. The most common circumstance is local extension of carcinoma of the cervix, endometrium, or rectum, or of metastatic tumours in the pouch of Douglas. In this case, destruction of the mucosa causes necrotic débris, altered and fresh blood and inflammatory exudate to appear in the sediment. Against this background, poorly differentiated cells with some squamoid features are found as a result of extension of cervical carcinoma, and cells with fragile cytoplasm and large, eccentric vesicular nuclei as a result of endometrial or rectal adenocarcinomatous invasion.

Less frequently metastases to the kidneys from various sites may occur, with or without representative cells in the urine. A classical example is that of a bronchogenic carcinoma, with multiple metastases to the kidneys and other organs. Cells consistent with small cell carcinoma may be found in bladder urine sediment and in ureteral catheter specimens from both sides. The appearance of the cells depends upon the source and cell type of the original tumour.

Body Fluids Cytology

The analysis of the cellular components of fluids from the thoracic and abdominal cavity has long been accepted as an important diagnostic procedure. It is rapid, inexpensive, and relatively painless. It is often used as an alternative to surgical exploration and approaches the same degree of diagnostic accuracy. Cytopathology is mainly concerned with the diagnosis of malignant from non-malignant effusions. In addition, those with experience in cytodiagnosis attain some skill in determining the type of neoplasm concerned.

Since the finding of malignant cells in body fluids usually connotes an advanced stage of disease and since therapeutic modalities for such disease carry a calculated risk, great restraint must be exercised if the cytologic evidence is not conclusive. It should also be remembered that patients with cancer may have effusions due to other causes.

The processing of fluids in the laboratory is an important step in ensuring accurate evaluation. Depending on the volume of the specimen, processing may include smear of the sediment, the use of membrane filters, or the preparation of cell blocks, singly or in combination. A large volume, delivered and processed immediately, is desirable.

BENIGN CELLS IN BODY FLUIDS

The cellular content of benign effusions varies with the etiological factors involved. Inflammatory exudates contain a preponderance of leucocytes, with a predominant cell type dependent upon the etiological agent, the chronicity of the infection and other factors.

Transudates in heart failure, cirrhosis and other less common conditions may present cytodiagnostic problems, especially where they are of long standing and accompanied by mesothelial proliferation.

Mesothelial cells usually occur singly, but they often appear as monolayer groupings of several cells. The central nuclei are regular, round or oval and have a uniform dispersion of chromatin. Chromocentres are prominent and one or more small, regular nucleoli may be seen. The nuclei occupy from one-third to one-half of the cell area. Binucleation is common and there is often a light area between the nuclei. The cytoplasm is usually basophilic but may be amphophilic. At the periphery of the cell there is frequently an area which appears less dense, giving a blurred outline to the cell. Vacuolization may reach extreme degrees, often displacing the normally centrally situated nucleus to the periphery. In these cells the nucleus appears moulded by the vacuoles and often presents a problem of differential diagnosis from adenocarcinoma. The mesothelial cells may contain phagocytosed cells or particulate matter.

The other cellular elements present consist of histiocytes, lymphocytes and, at times, red blood cells.

MALIGNANT CELLS IN BODY FLUIDS

Malignant effusions are usually caused by metastatic carcinoma and are much more common in females, with breast carcinoma accounting for over half of the pleural effusions and ovarian carcinoma for almost half of the ascites specimens in most places. In males the majority of malignant pleural effusions are caused by lung carcinoma, and about half of the ascites cases by pancreas and stomach cancer.

A detailed description of cells derived from the various types of carcinoma is given elsewhere in the text and will not be repeated here.

Mesotheliomas are relatively rare tumours and may present complicated differential diagnostic problems between metastatic adenocarcinoma on the one hand and benign mesothelial proliferation on the other.

The cells and nuclei are greatly enlarged. The nuclei are irregular and the chromatin often is clumped, with a tendency to margination. The cells appear both singly and in mulberry clusters. The clusters often appear as three-dimensional balls, unlike the usual monolayer of the benign proliferations.

Chromocentres are prominent and large eosinophilic nucleoli, sometimes with irregular outlines, are found in a large proportion of the cells. In the clumps there may be considerable moulding of the nuclei. The large cytoplasmic vacuoles described in the benign mesothelial proliferations are not usually present. There are no other cytoplasmic features to aid in the differential diagnosis.

Gastro-oesophageal Cytology

Gastric cancer is a significant, and in some populations the predominant, cancer causing death. The survival rate is extremely poor and has remained so over the past 60 years in spite of improved surgery, irradiation and chemotherapy.

The incidence and mortality rates for gastric cancer are dropping in many countries, but are still substantial in some countries, such as Chile, Czechoslovakia, Finland, Iceland, and Japan. The disease therefore constitutes a considerable problem and early detection methods have been used in an attempt to improve survival rates.

Cytology is now considered one of the best methods for diagnosing gastric cancer in its earliest stages, especially since the gastrofibrescope has made possible selective brushing or washing of the lesion.

The histological examination of the gastrectomy specimen must be thorough and comprehensive enough to avoid missing a lesion that is difficult to visualize. Ideally, the partly excised organ should be examined before fixation, to identify any suspicious areas present in addition to any obvious lesion such as an ulcer. Several sections for examination should be taken from several areas of the stomach. It has been asserted that up to 10% of stomachs with benign ulcers demonstrate surface cancers, often quite small, in areas well away from the ulcer.

Attention must be given to extragastric tumour cells which may be found in gastric specimens, especially in lavage samples. They may derive from swallowed saliva or sputum, in which case the characteristic tumour patterns seen in those specimens prevail, modified only by dilution and/or digestion.

The incidence of oesophageal carcinoma in some countries is greater than that of gastric cancer. Moreover, the prognosis is worse. There is thus a great need to achieve early detection for such lesions, and cytology plays an important role in this field, especially now that improved oesophagofibrescopes are being used for sampling under direct vision.

OESOPHAGUS

BENIGN CELLS IN OESOPHAGEAL SPECIMENS

1. *Normal cells*

The normal squamous cells derived from the oesophagus are usually of the intermediate type, with pale vesicular nuclei and non-granular cytoplasm staining either eosinophilic or cyanophilic. The presence of superficial squames and often squamous cells from other layers seen in gastric and oesophageal washings may come from the oral pharynx or buccal cavity. The background pattern is clear and devoid of débris but may be complicated by swallowed sputum.

2. *Inflammatory and degenerative cell changes* (oesophagitis and ulceration)

There is enlargement of nuclei, with granular changes in the nucleoplasm and prominent chromocentres. Sometimes there is a moderate to marked degree of orangeophilia (Papanicolaou stain), with some keratinized and anucleate squames present. This is accompanied by inflammatory exudate.

MALIGNANT CELLS IN OESOPHAGEAL SPECIMENS

1. *Squamous cell carcinoma*

The classical features of squamous cell carcinoma are displayed in the cytological specimens, with a pleomorphic picture of malignant squamous cells, blood, and débris. Keratinized, degenerate, and anucleate squames are often present.

2. *Adenocarcinoma*

Oesophageal adenocarcinoma has the same cellular pattern as adenocarcinoma of the stomach.

STOMACH

BENIGN CELLS IN GASTRIC SPECIMENS

1. *Normal cells*

It is unlikely that distinctive identification can be made of the different types of epithelial cells composing the gastric mucosa, such as the chief, parietal, accessory and mucus neck cells and this is in any case not essential in the diagnosis of malignancy.

The normal gastric mucosal cell is columnar and occurs either singly or in sheets. The nucleus is round or oval, with finely granular chromatin and a thin nuclear rim. It is positioned at the lower pole when seen in the side view, and eccentrically placed when viewed end-on.

The smear background in lavage is clean but mucoid and may contain very few gastric cells. It may be contaminated by material from the buccal mucosa or respiratory tract.

2. *Inflammatory and degenerative cell changes*

Inflammatory changes in the gastric mucosal cells are characterized by varying degrees of enlargement and diffuse hyperchromasia of the nucleus, with prominence of the nuclear membrane. Some enlargement of nucleoli may also occur.

The whole cells may increase in size, with granular or floccular change in the cytoplasm, often to the extent of vacuolation, and this may be perinuclear in position. Some variation in stain uptake may also be present.

Degenerative changes may take the form of nuclear enlargement, due to hydropic degeneration or contraction and pyknosis or fragmentation (karyorrhexis), or there may only be some degree of irregularity of the nuclear membrane. The cytoplasm may undergo marked vacuolation, with some loss of cytoplasmic borders or complete stripping of the nucleus.

3. *Atrophic gastritis and intestinal metaplasia*

In cases of hypochlorhydria, atrophic gastritis and gastric ulcer a characteristic picture is present, with atrophic, inflammatory and degenerative changes occurring in epithelial cells, accompanied by a characteristic exudate. The latter consists of blood, leucocytes, histiocytes, lymphocytes and plasma cells. There may be proliferation of organisms, both bacterial and fungal.

The presence of intestinal metaplasia, a common accompaniment of chronic atrophic gastritis and achlorhydria, occurs with increasing frequency as age advances. It occurs in a high proportion of stomachs bearing cancers. It does, however, also occur to some degree in benign ulcer-bearing stomachs. As cancer may develop in stomachs displaying atrophic gastritis and intestinal metaplasia, it is important that these entities should be recognized cytologically.

The characteristic changes of metaplasia are: the presence of enlarged, often globular, cells containing mucus, which in turn gives a foamy appearance; at the luminal end of the cell there may be an optically dense bar; the nucleus is eccentrically positioned, which in an extreme degree results in a benign signet-ring form, sometimes causing diagnostic difficulty.

Another type of intestinal metaplasia is seen as hypertrophy of the mucosa in small areas. This is found in cases of atrophic gastritis and occasionally surrounding small foci of carcinoma in the antrum. Cytologically, the cells are slender and elongated and exfoliate in bundles. The appearance is essentially benign.

4. *Regenerative cell changes*

There may be uniform but increased activity and/or enlargement of nuclei, with prominent nucleoli in cells, characteristically in sheets or clusters. In some there is a prominent but evenly thickened nuclear membrane, with a moderate increase in granular or reticular chromatin. There may be some nuclear crowding and loss of cytoplasmic boundaries, with an occasionally moderate increase in the nuclear/cytoplasmic ratio.

5. *Borderline lesions*

These show greater degrees of nuclear change and crowding than occurs in regenerative states but appear to be short of frank cancer. The greater tendency to anisocytosis, hyperchromatism and anisokaryosis causes difficulty in interpreting these lesions. Such cases should be carefully followed up.

MALIGNANT CELLS IN GASTRIC SPECIMENS

Adenocarcinoma

The usual characteristic changes of glandular cancer are seen in varying degrees of cell differentiation. The cellular appearance depends on the method of collection, finer preservation being possible with brush or biopsy techniques than in the lavage specimen.

The nuclei show marked and unequal enlargement, with a variation in shape and position, often at the extreme margin of the cells. The chromatin is coarse and unequally clumped, with sometimes an irregularly thickened membrane and often marked variation in the stain uptake, even in the same cluster. Hyperchromatism is not always a distinctive feature. The nucleoli are large and prominent, and often multiple, and the increased nucleolar/nuclear ratio provides a better diagnostic criterion than the nuclear/cytoplasmic ratio, since mucus secretion or loss of cytoplasmic borders make the latter ratio unreliable.

The less well differentiated forms of adenocarcinoma display a coarser chromatin pattern; great crowding of nuclei with lesser amounts of cytoplasm often result in stripped nuclei. There is also a great tendency to cell disaggregation.

The background pattern is one of atrophic gastritis with an increased amount of fresh and old blood and cell débris.

Carcinoma in situ (intra-epithelial carcinoma) is an entity that is not sufficiently established histologically to be accepted by all and, in any event, the cytological picture is similar if not identical to that of gross adenocarcinoma of the stomach.

Surface carcinoma, in which the neoplastic process is confined to the surface layers of mucosa (intramucosal cancer) or submucosa (submucosal cancer) should be separated from the more advanced tumours. These surface lesions may present with a relatively clean smear pattern, with fewer degenerative changes than are seen in more gross tumours. Nevertheless, cytologically there is no distinctive pattern to identify lesions limited to the mucosal surface.

MISCELLANEOUS TUMOURS [1]

Lymphosarcoma cells tend to disaggregate, though small sheets or clusters do occur. The primary cell type is round or oval, with a rounded,

---

[1] The cells of the lymphoreticular series are best displayed by the Giemsa stain.

fairly densely staining nucleus displaying irregular chromatin condensation and an irregular, though poorly defined, nuclear border. The nucleoli are not prominent but are often multiple and irregular. The cytoplasm is deep blue and contains numerous small vacuoles.

Reticulosarcomas are more segregated and present larger, rounded and more pale-staining cells than are seen in lymphosarcoma.

Well-preserved cells have abundant cytoplasm staining pale or dark blue, sometimes with many small vacuoles. The nuclear chromatin is finely granular, with large prominent nucleoli of varying numbers and size. The nuclear border is thin and often indefinite in Giemsa-stained material but more prominent when stained by the Papanicolaou technique.

Hodgkin's disease is rare in the stomach and displays a more pleomorphic picture, with immature reticulum cells and the characteristic Reed-Sternberg cells. The latter possess minimal pale-staining cytoplasm and two or more vesicular nuclei, often with a mirror image, and prominent nucleoli. There is a background pattern of neutrophils, eosinophils, histiocytes and lymphocytes.

Other sarcomas are extremely rare, the least rare among them being leiomyosarcoma. The cells are irregular, with a reticular nuclear pattern and numerous prominent nucleoli. The cytoplasm is pale-staining, variable in amount, with indistinct margins. In all sarcomatous tumours a background similar to carcinoma occurs.

SECONDARY TUMOURS

These are rare and will not be considered here.

Fine Needle Aspiration Cytology

Aspiration cytology is a method of cell investigation based on the removal of a cell sample by means of a needle attached to a syringe.

It should be performed by a physician with a great deal of experience, whether a cytopathologist or clinician. In some institutions it has been found that the aspiration procedure is best performed in a single clinical centre by a cytopathologist.

The indications for fine needle aspiration are primarily situations where there is clinical suspicion of a malignant tumour. After experience has been gained the procedure may be found useful in other clinical situations.

A long period of initial experience is required before the method may be applied as a reliable diagnostic procedure in the management of malignant disease. A great deal of knowledge of tumour pathology and histology is required before the method can be learned. Expertise in aspiration cytology may be gained by a period of study in an established centre. A minimum period of six months of study is recommended for pathologists.

It has been documented that the so-called " fine needle " method gives optimal smears and is less traumatic to the patients than large-calibre needles which are still in use in some centres.

Virtually any organ of the body can be investigated by aspiration cytology. It is not possible within the framework of this volume to illustrate all aspects of aspiration cytology; therefore only a few of the more widely used applications will be discussed.

THYROID

Aspiration of palpable lesions of the thyroid is useful to distinguish thyroiditis, hyperplasia and neoplasia. In thyroiditis inflammatory cells are present in addition to benign epithelial cells. In hyperplasia, epithelial cells are surrounded by colloid. The absence of colloid and a high degree of cellularity are the characteristic features of neoplasia.

THYROIDITIS

Three different types of thyroiditis are usually distinguished:

Acute thyroiditis is characterized by the presence of granulocytes in a protein-rich background. A word of caution is necessary: aspirates from giant cell undifferentiated carcinoma or squamous cell carcinoma may also contain numerous granulocytes.

Subacute thyroiditis is characterized by multinucleated giant cells intermingled with lymphocytes and macrophages. Follicular cells often show signs of degeneration.

Chronic lymphoid thyroiditis is characterized by the presence of numerous lymphocytic cells with a scanty admixture of follicular cells. Follicular cells may show enlarged hyperchromatic nuclei surrounded by eosinophilic granular cytoplasm.

HYPERPLASIA

The aspirates show variable features. Some consist of fluid containing degenerated blood cells, foamy phagocytes and degenerated follicular cells. Others consist of amber-coloured colloid with a scant admixture of follicular cells. Aspirates taken from the cellular areas will consist almost exclusively of follicular cells, and cytologically the lesions may not be distinguishable from true neoplasia.

ADENOMAS

The aspirates from adenomas are characterized by high cellularity in the absence of colloid. The neoplastic cells are usually arranged in follicular structures. It should be pointed out that in aspiration smears adenomas cannot be reliably distinguished from follicular carcinoma. The cytologist should only report a follicular neoplasm and the decision on its malignancy should await the histological examination.

CARCINOMAS

Histologically, thyroid carcinomas are classified as follows:

Follicular carcinoma. Aspirates contain clusters of epithelial cells, often imitating follicular structures in the absence of colloid. Cytologically, follicular carcinomas cannot be reliably distinguished from follicular adenomas.

Papillary carcinoma. Aspirates contain clusters of epithelial cells, often in monolayers. Some cells show cytoplasmic inclusions in the nucleus. Degenerated carcinoma cells are often present together with phagocytes. Multinucleated histiocytes can also be observed.

Medullary carcinoma. The carcinoma cells occur singly or in clusters. No follicular or papillary structures are seen. The cells are oval or round and a varying number may exhibit granulation of the cytoplasm. Amorphous material consisting of amyloid may be present in the smear.

Squamous cell carcinoma. Primary squamous cell carcinoma is rare. When squamous carcinoma cells are aspirated, the possibility of extension from a cancer in the larynx, trachea or oesophagus must be considered.

Undifferentiated (anaplastic) carcinoma :

(a) *Giant cell type.* These tumours are composed of various proportions of giant cells. Necrotic material and numerous inflammatory cells are present.

(b) *Small cell type.* These tumours are composed of a rather monomorphic population of oval-shaped cells which cytologically can be readily distinguished from malignant lymphoma.

BREAST

It is possible by means of aspiration cytology to differentiate between the benign and the malignant lesions of the breast.

BENIGN CELLS IN BREAST ASPIRATES

It is not possible to establish an exact cytologic diagnosis of most benign lesions, except for cysts. Knowledge of the clinical presentation and the results of mammography and galactography are helpful in the interpretation of the aspiration cytology.

Inflammatory lesions of the breast accompanied by necrosis and fibrinous exudate should be interpreted in a most conservative manner; endothelial cells, atypical ductal cells and histiocytes may be mistaken for cancer cells.

Breast cysts

If the aspirated fluid is clear, the sediment contains, as a rule, only a few foam cells. If the fluid is cloudy, foam cells, inflammatory cells or ductal cells may be found. Ductal cells appear singly or in clusters. Some of them have eosinophilic cytoplasm (apocrine cells). Ductal cells usually show signs of degeneration.

Other benign epithelial lesions

Papillomas, fibroadenomas, and various forms of proliferative mastopathy such as sclerosing (fibrosing) adenosis all have a similar cytologic presentation, except for quantitative differences in the number of epithelial cells. The cells occur predominantly in cohesive clusters with some overlapping. The cells have small round or oval nuclei and variable amounts of cytoplasm. In addition, naked (single) nuclei are observed.

In fibroadenomas and papillomas the epithelial cells are sometimes very numerous and they may be arranged in monolayers. Fragments of loose connective tissue stroma may occasionally be observed in aspirates from fibroadenomas.

MALIGNANT CELLS IN BREAST ASPIRATES

The cytologic diagnosis of breast cancer must be based on unequivocal evidence if treatment without surgical biopsy is contemplated. In debatable cases, when the cytologic evidence of cancer is scanty, histologic diagnosis is mandatory before treatment.

Carcinoma

Infiltrating duct carcinoma, comprising the majority of all breast cancers, is characterized principally by clusters of cancer cells that vary in size when compared with normal ductal cells. A small proportion of these tumours yield very small cancer cells, comparable in size to normal ductal cells. In the majority, however, the cells are significantly larger.

Regardless of size of the cells the clusters are characterized by a loose arrangement of superimposed cells, surrounded by many single cancer cells with well-preserved cytoplasmic borders. The nuclei, which are best observed in the single cells, are usually hyperchromatic and large, and are occasionally located eccentrically. The use of various fixation and staining techniques may account for significant differences in the appearance of cancer cells.

The greatest diagnostic difficulty may be encountered with small cancer cells, particularly when the material is scanty. The latter situation may occur with infiltrating duct carcinomas of the scirrhous type.

Intraductal carcinoma that may be incidentally aspirated has a similar cytologic presentation, except for the presence of cell degeneration and necrosis.

A variant of infiltrating duct carcinoma, the so-called *apocrine cell carcinoma* is characterized by large cancer cells with eccentric nuclei. The cytoplasm contains eosinophilic granules when stained with haematoxylin and eosin or with Papanicolaou stain.

Infiltrating lobular carcinoma. Occasionally, the cancer cells in the smears of a breast carcinoma occur singly and do not form clusters. Most of such cases are infiltrating lobular carcinomas but a few are ductal carcinomas.

Medullary carcinoma. Occasionally, in addition to cancer cells, numerous lymphocytes are found in the smears. Such smears may reflect a medullary carcinoma with lymphoid stroma—a tumour with a better prognosis than most breast cancers. However, the same cytologic presentation may occur when a lymph node with metastatic carcinoma is aspirated. This differential diagnosis must be based on clinical presentation.

Mucinous (colloid) carcinoma. Aspiration biopsy yields mucus-producing cancer cells surrounded by pools of mucus appearing as a pale homogeneous substance. The cells are uniform in size and appearance and have a round, eccentric nucleus and abundant pale cytoplasm.

Other rare variants of breast cancer such as papillary, squamous and adenoid cystic may be occasionally identified in smears.

Sarcoma

A variety of sarcomas may be observed in the breast, such as sarcomas arising in cellular intracanalicular fibroadenomas (malignant " cystosarcoma phyllodes "), stromal sarcoma, angiosarcoma, liposarcoma or malignant lymphoma. The most common is malignant " cystosarcoma phyllodes ", which may be identified in smears by the presence of fragments of stroma composed of loosely arranged bizarre fibroblast-like cells.

PROSTATE

Fine needle aspiration of the prostate can be performed with a perineal or transrectal approach. The latter offers better and more accurate sampling with less discomfort to the patient. A long needle with a needle guide is required. Because of the possibility of infection, it is recommended that the transrectal method be used only in situations where prostatic cancer is suspected.

BENIGN CELLS IN PROSTATE ASPIRATES

Smears from a normal prostate consist of cohesive clusters of epithelial cells of equal size and appearance. Acinus-like structures are rare.

Prostatitis, particularly the granulomatous form with eosinophilia, may produce cellular atypias that may be confused with cancer by an inexperienced observer and may require further investigation by repeat aspiration or other means.

A word of caution is required: incidental aspiration of seminal vesicles may harvest epithelial cells characterized by nuclear atypia. Such cells may be identified by dark brown cytoplasmic pigment. Phagocytosis of sperm may also be observed.

MALIGNANT CELLS IN PROSTATE ASPIRATES

From the practical point of view it may be advisable to separate prostatic adenocarcinomas into three groups which may call for different approaches to treatment:

1. The *well differentiated* prostatic carcinoma is characterized by a pattern of atypical acini, lined by cells with slightly enlarged and hyperchromatic nuclei in which a nucleolus can be clearly visualized.

2. *Moderately well differentiated* prostatic carcinoma shows a greater

degree of nuclear abnormality, although acini may still be present. In addition, numerous clusters of cancer cells are observed.

3. In *poorly differentiated* prostatic carcinoma the cytologic pattern is dominated by loosely arranged clusters and single, clearly identifiable large cancer cells. Markedly abnormal single cells may dominate the picture.

Infrequently, other primary or secondary cancers of the prostate may be diagnosed by aspiration biopsy, such as rhabdomyosarcoma, seen mainly in children, or other rare tumours. The most frequently seen secondary cancer is an extension of bladder cancer invading the prostate. Metastatic tumours such as malignant lymphoma may also be seen.

EFFECTS OF TREATMENT

Treatment with oestrogens is reflected in the occurrence of metaplastic squamous cells.

LYMPH NODES

Aspiration biopsy of lymph nodes is a widely used method in the diagnosis of clinically abnormal lymph nodes. The method can be applied in the diagnosis of malignant lymphomas and related disorders and in the diagnosis of metastatic cancer.

If possible, the aspiration should be performed on more than one lymph node. Prior to the evaluation of aspirated material, it is essential that details of the history, clinical presentation of the nodes, and haematological work-up be known to the cytopathologist. This provides additional insurance against errors in what is a very difficult field of cytologic diagnosis.

BENIGN CELLS IN LYMPH NODE ASPIRATES

Aspiration smears of benign hyperplastic lymph nodes usually yield a variable population of cells: lymphocytes in various stages of maturation, stem cells and reticulum cells (histiocytes). Polymorphonuclear leucocytes and plasma cells may also be present.

If the aspirate yields only one cell population (for example, lymphocytes), great caution must be exercised in interpretation as there may be confusion with malignant lymphoma.

Many of the granulomatous inflammatory diseases, notably tuberculosis, may yield necrotic material, giant cells and epithelioid cells. The limitations of the diagnosis of granulomatous disease on the basis of aspiration smear must be kept in mind; similar changes may also occur in the presence of metastatic cancers where necrosis is present.

MALIGNANT CELLS IN LYMPH NODE ASPIRATES

Lymphosarcomas (malignant lymphomas) and related disorders

These neoplasms form a spectrum of disease that varies from the lymphocytic lymphosarcomas (well-differentiated lymphocytic lymphoma), composed predominantly of lymphocytes, to neoplasms composed of obviously malignant large cells, such as lymphoblastic lymphosarcoma or reticulosarcoma. Intermediate forms may be observed.

Lymphocytic lymphosarcoma (lymphocytic lymphoma). The cytologic identification of the well-differentiated lymphocytic lymphoma may be extremely difficult if the smear shows only lymphocytes of variable sizes. In poorly differentiated lymphomas the diagnosis is more readily made. In many instances a histologic confirmation of disease is required.

Lymphoblastic lymphosarcoma (stem cell lymphoma) is characterized by large cells with scanty cytoplasm. The nuclei are nearly round, are hyperchromatic, and contain prominent nucleoli.

Reticulosarcoma (histiocytic lymphoma). The cells are relatively easy to identify; they have large hyperchromatic, irregular nuclei with large, often multiple, nucleoli, and variable amounts of cytoplasm. They may appear singly or in clusters. In the majority of cases lymphocytic cells of various stages of maturation can also be recognized in the smears. The differential diagnosis of reticulosarcoma comprises metastatic tumours of small cell size, such as nasopharyngeal carcinoma, oat-cell carcinoma and, in children, neuroblastomas, Ewing's sarcoma, etc.

Hodgkin's disease. Aspirates from lymph nodes affected by Hodgkin's disease show a variety of cell patterns, depending on the histologic type. The diagnosis can be made only if clearly identifiable large binucleated or multinucleated Reed-Sternberg cells are observed, which must contain large nucleoli.

A more detailed classification will be found elsewhere.[1]

Metastatic tumour cells in lymph node aspirates

The knowledge of prior history of cancer is very helpful in the evaluation of lymph node aspirates. In the absence of such a history, the anatomical location of the lymph node containing metastatic cancer may be extremely helpful in the identification of the primary tumour.

A world of caution is necessary: in certain situations, aspirates may contain epithelial cells that mimic metastatic cancer; for example, branchial cleft cysts or carotid body tumours may be mistaken clinically for enlarged lymph nodes. The aspirates from branchial cleft cysts may be mistaken for

---

[1] Mathé, G., Rappaport, H., O'Conor G. T. & Torloni, H. *Histological and cytological typing of neoplastic diseases of haematopoietic and lymphoid tissues.* Geneva, World Health Organization, 1976 (*International Histological Classification of Tumours*, No. 14).

squamous carcinoma and those from a carotid body tumour may mimic a variety of malignant tumours, including metastatic thyroid carcinoma.

The cytology of aspirates of lymph nodes with metastatic tumours is extremely variable and depends on the histologic patterns of the primary tumours. Certain tumour types are easily recognizable, such as squamous keratinizing carcinoma, adenocarcinoma and melanoma. An experienced worker can identify the site of origin of some metastatic adenocarcinomas, such as thyroid or kidney.

Annex 1

URINE AND PROSTATIC SECRETIONS

Methodology

One or several specimens of urine are obtained on consecutive days (by catheter in women) from a well-hydrated patient who should, if possible, move about actively. A midstream specimen is of no use. If the entire specimen cannot be sent, the first and last portions are more suitable for diagnosis. The fresh urine should be sent immediately to the laboratory. If the specimen cannot be delivered to the laboratory within a short time, an equal volume of 50% alcohol must be added.

A. *Centrifugation*

1. A drop of bovine albumin can be added to fresh urine to increase adhesion of cells to the slide. Alternatively, the slides should be lightly albuminized with egg albumin or pectin.

2. Centrifuge for 5–10 minutes.

3. Spread on 2–4 slides and fix immediately in ether-alcohol or 95% alcohol.

4. Stain by the Papanicolaou method.

5. Dehydrate, clear, and mount in a neutral synthetic medium.

6. A fat stain of air-dried smears of fresh urine sediment can be employed for the identification of cells from renal parenchymal carcinoma.

B. *Cytocentrifugation*

1. Well-mixed fresh urine is poured into special cuvettes.

2. The slides are albuminized.

3. The cells are directly centrifuged on to the slide. The time of centrifugation must be shortened if the urine contains only a few cells.

4. Fixation, staining and further processing, as above.

C. *Filtration*

1. Fixation in equal parts of 10% formalin and 2% saponin, or in 50% alcohol.

2. With a syringe, push the mixture gently through a 25-mm filter, pore size 3–5 μm, in a Swinnex unit.

3. Stain by the EA 36 Papanicolaou method with the following modifications:

(*a*) The filter is placed immediately into 80% alcohol.

(*b*) The haematoxylin and EA 36 staining time is reduced with fresh stains.

(*c*) Wash by placing the filter in 3 changes of water for 2 minutes.

(*d*) The filter is placed on blotting paper with the cells facing upwards.

(*e*) Mount with thick Canada balsam on both sides of the filter.

(*f*) Seal the edges of the cover slip with nail varnish.

D. *Prostatic cytology*

1. In advanced or treated cases of prostatic cancer, the sclerotic tumour yields few, if any, cells. In cytologically suspicious cases, histologic verification should be obtained. Excessive manipulation of the tumour may enhance its spread.

2. When the patient is to have prostatic palpation or massage, slides and fixing solution should be at hand. In the case of massage, contamination of the specimen with seminal fluid should be avoided as far as possible.

3. Any secretions obtained should be spread on two or more slides and dropped at once into the fixing solution.

4. The first voided urine following the procedure should be collected, whether or not secretions were obtained. The sediment of this urine is prepared as described above.

Annex 2

GASTRO-OESOPHAGEAL SPECIMENS

Preparation of the patient

Overnight fasting with ample water intake is essential. If there are signs of pyloric obstruction or stasis, one or more evening gastric washes may be necessary on the day preceding the test. Local anaesthesia may be necessary, especially when the fibrescope is used. The patient should be sitting or in a semi-reclining position for oesophageal washings and either sitting or fully reclining for gastric washings.

Simple saline wash technique

The Levine tube is passed with direct aspirations and saline washes made at certain levels marked on the tube: throat (26 cm), oesophagus (36 cm), cardia (45–52 cm), body and pylorus (60–65 cm) and duodenum (75 cm).

Washings of the oesophagus

These can be made at varying levels from 26 to 50 cm with small volumes of saline (20–40 ml) and immediate aspiration, if possible. A larger volume of fluid is then passed into the upper oesophagus and finally aspirated from the stomach.

Gastric washings

Gastric washings are performed with a similar tube. Approximately 200–300 ml of isotonic saline is introduced into the upper part of the stomach. The patient is then exercised by bending and turning and the abdomen is massaged or ballotted to ensure thorough washing of the stomach surface. A degree of internal ballottement is then carried out by means of withdrawing and returning 30–50 ml of fluid with a syringe or pump. This sample is fully aspirated and the procedure repeated once more, so that two samples are obtained.

The entire volume of the remaining fluid and wash sample(s) is rapidly chilled and centrifuged. From 8 to 10 smears are prepared and fixed in equal parts of ether and 95% alcohol, or 95% alcohol.

Screening must be performed by a highly experienced cytoscreener.

The oesophagofibrescope technique

This is ideally an end-viewing instrument with brush and biopsy attachments. It is introduced with the patient lying in the left lateral position. Selective brushings or biopsy samples are taken of the suspect sites or lesions under direct vision.

As haemorrhage always occurs with biopsy, the brush sample is collected first; sometimes biopsy is not done. More than one brush sample may be taken with the fibrescope still in position.

The brush samples are smeared, then the brush is washed in saline, which is centrifuged, and smears are made. The biopsy material is lightly touched or imprinted on clean dry glass slides which are appropriately fixed, depending on the stain used (dry smears for Giemsa and immediate wet fixation for Papanicolaou stain). The biopsy fragments are then fixed for histological sectioning.

The gastrofibrescope

This is ideally a side-viewing instrument with brush and biopsy attachments, and the same channel is used to introduce the narrow vinyl tube for flushing the lesion under direct visualization.

Initial passage of the tube is performed with the patient in the left lateral position. Selective brushing and biopsy and/or lavage is performed, and the material prepared in the same manner as for the oesophagus.

Annex 3

TECHNIQUE OF ASPIRATION BIOPSY OF PALPABLE LESIONS

Aspiration biopsy is performed to disclose the nature of lesions detected by palpation, X-ray examination, scintiscan, or other methods.

Instruments

Apart from a syringe and a well-fitting needle, aspiration biopsy requires little equipment. The syringe should have a special handle which permits a one-hand grip while biopsy is being performed. Fine needles (22-gauge, outer diameter about 0.6 mm) are recommended; they reduce the risk of trauma and minimize the admixture of blood. Occasionally, however, a thicker needle is required if the initial puncture suggests that fibrotic tissue is present, or for aspiration of cystic contents. Larger bore needles equipped with a stilette are required for aspiration biopsy of skeletal foci.

For transrectal or transvaginal aspiration biopsy, a needle guide is helpful. This can be simply a piece of plastic tubing fixed to the gloved palpating finger with adhesive tape and a finger-cot pulled over it. A special instrument for this type of aspiration biopsy was devised by Franzen et al. It consists of a fine metal tube having on its distal end a steering ring for the palpating finger. The ring is open on its dorsal aspect and can be adjusted to fit the operator's finger. To facilitate the introduction of the needle into the guide, the latter is enlarged at its proximal end in funnel fashion. An adjustable plate midway along the guide supports the instrument, resting on the thenar eminence during needling. The needle is of flexible stainless steel, approximately 22 mm long and 22-gauge. Its proximal end is somewhat thickened. The transition from the thicker to the thinner part of the needle serves as a marker.

The syringe and the needles must be absolutely dry during the biopsy, otherwise reading the slides is difficult or impossible owing to osmotic cytolysis.

The puncture (transcutaneous aspiration biopsy of palpable lesions)

The skin is wiped with an antiseptic solution and the suspect lesion is held with one hand in a position favourable for needle biopsy. No anaesthesia is required. Preliminary nicking of the skin in order to avoid contamination of the aspirate with squamous epithelium or other cells from around the needle track is unnecessary when a fine needle is used.

When the needle has entered the tumour area, the piston of the syringe is retracted, thus creating a vacuum, while the needle is guided in a straight line through the lesion. In order to obtain sufficient material, particularly from fibrotic lesions, the needle may have to be moved back and forth three or more times and directed into different areas of the tumour. Throughout this manipulation, negative pressure is maintained in the syringe by keeping the piston retracted.

When the aspiration has been completed, the piston is released so that the pressure in the syringe equalizes before the needle is withdrawn from the lesion.

After the needle has been withdrawn, the syringe is disconnected, filled with air and reconnected. The material in the needle is expelled on to a glass slide, care being taken to deposit it as a single drop. The syringe and needle are then washed with saline, which is centrifuged and smears are made.

Preparation of smear

The drop of aspirate is spread on the slide and *immediately* immersed in 95% ethyl alcohol or alcohol-ether in equal parts. After fixation it is stained by the usual Papanicalaou technique. In addition, air-dried smears may be made and stained with any of the usual haematological stains. When larger bore needles are used, particles of aspirate may be placed in fixative for sectioning.

INDEX

	Page	Figures
PULMONARY CYTOLOGY	19	
Adenocarcinoma	27	58–77
bronchogenic	27	58–68
bronchiolo-alveolar	27	69–77
Asbestosis, *see* Ferruginous bodies		
Atypical squamous metaplasia	23	18–27
mild	23	18–21
moderate	23	22–24
severe	23	25–27
Contaminants	28	95
Curschmann's spiral	28	96
Dysplasia, *see* Atypical squamous metaplasia		
Epithelium, normal	22	1–6
Ferruginous bodies	29	99
Herpes, *see* Viral changes		
Large cell carcinoma	28	78–84
Macrophages	24	8, 9
Metaplasia, *see* Atypical squamous metaplasia and Squamous metaplasia, benign		
Small cell carcinoma	26	49–57
Squamous cell carcinoma	25	28–48
in situ	25	28–39
invasive	26	40–48
Squamous metaplasia, benign	23	16, 17
Tuberculosis (granuloma, epithelioid cells)	25	10–12
Vegetable cells, *see* Contaminants		
Viral changes	28	13–15
URINARY TRACT CYTOLOGY	30	
Adenocarcinoma (bladder)	34	136, 137
Adenocarcinoma (prostate)	35	141–143
Carcinoma in situ (bladder)	33	129, 130
Cystitis		
acute	32	106
chronic	33	107
Cystitis glandularis, von Brunn's nest	32	108, 109
Inflammation	32	—
Malakoplakia	32	—
Metaplasia, *see* Squamous metaplasia		
Metastatic tumours	36	148–151
Non-papillary carcinoma, *see* Urothelial carcinoma		
Papillary carcinoma, *see* Urothelial carcinoma		
Papillary tumour	—	113–116

	Page	Figures
URINARY TRACT CYTOLOGY (Continued)		
Renal parenchymal carcinoma	35	144–147
		152, 153
Rhabdomyosarcoma	35	138, 139
Sarcoma	35	—
Schistosomiasis	—	135
Seminal vesicle cell	48	111
Squamous cell carcinoma	34	133, 134
large cell non-keratinizing	34	—
keratinizing	34	—
Squamous metaplasia	32	102–105
non-keratinizing	32	102, 103
keratinizing	32	104, 105
Transitional cell tumours, *see* Urothelial carcinoma and Squamous cell carcinoma		
Undifferentiated carcinoma	35	123–126
Urothelial carcinoma	33, 34	117–122
papillary	34	117–120
non-papillary	34	121, 122
Urothelium, normal	31	100, 101
BODY FLUIDS CYTOLOGY	37	
Benign cells	37	157–161
Malignant metastatic cells	—	166–190
Mesothelioma	38	162–165
GASTRO-OESOPHAGEAL CYTOLOGY	39	
Adenocarcinoma (oesophagus)	40	—
Adenocarcinoma (stomach)	42	244–254
mucinous (signet-ring type)	—	255–264
papillary	—	240–243
Atrophic gastritis, *see* Gastritis		
Borderline lesions (stomach)	41	—
Carcinoma in situ (oesophagus)	—	194–196
Gastric mucosa, normal	40	201–204
Gastritis, atrophic	41	205–209
Hodgkin's disease (stomach)	43	273, 274
Inflammation (gastric mucosa)	40	204
Intra-epithelial carcinoma, *see* Carcinoma in situ		
Leiomyosarcoma (stomach)	43	268–270
Lymphosarcoma (stomach)	42	271
Metaplasia, intestinal	41	210–213
Metastatic tumours (stomach)	42	—
Mucinous adenocarcinoma, *see* Adenocarcinoma (stomach)		
Oesophagitis	38	193
Papillary adenocarcinoma, *see* Adenocarcinoma (stomach)		
Polyp, adenomatous	—	228–232
Polyp, malignant	—	233–239
Regenerative changes in gastric mucosa	41	214, 215
Reticulosarcoma (stomach)	43	272
Sarcoma (stomach)	42, 43	268–274
Squamous cell carcinoma (oesophagus)	40	199, 200

CYTOLOGY OF NON-GYNAECOLOGICAL SITES

	Page	Figures

GASTRO-OESOPHAGEAL CYTOLOGY (Continued)

	Page	Figures
Squamous epithelium, normal (oesophagus)	39	191, 192
Surface carcinoma (stomach)	42	218–227

FINE NEEDLE ASPIRATION CYTOLOGY 44

	Page	Figures
Thyroid	44–46	275–295
Breast	46–48	296–340
Prostate	48, 49	341–358
Lymph nodes	49–51	359–379
Adenocarcinoma of the prostate	48, 49	346–353
moderately well differentiated	48	349, 350
poorly differentiated	49	351, 352
well differentiated	48	346–348, 353
Adenoma (thyroid)	45	281–283
Apocrine cells (breast)	46	307–309
Breast, normal and non-malignant	46	296–318, 322–324
Carcinoma of the breast	47, 48	325–340
apocrine cell	47	334–336
duct, infiltrating	47	325–327
intraductal	47	—
lobular	47	328–330
medullary	47	331–333
mucinous (colloid)	47	337–340
Carcinoma of the thyroid	45	284–295
follicular	45	284–286
medullary	45	290–292
papillary	45	287–289
squamous cell	45	—
undifferentiated (anaplastic)	46	293–295
giant cell	46	293–295
small cell	46	—
Cyst (breast)	46	312–314
Fibroadenoma (breast)	46	315–318
Hodgkin's disease (lymph node)	50	367–370
Hyperplasia (lymph node)	49	359, 360
Hyperplasia (thyroid)	45	275–277
Lymphoma		
histiocytic, see Reticulosarcoma		
lymphoblastic, see Lymphosarcoma		
stem cell, see Lymphosarcoma		
Lymphosarcoma	50	361–364
lymphoblastic	50	—
lymphocytic	50	361–364
Mastopathy, proliferative	46	—
Metaplasia, squamous (prostate)	49	353
Metastatic carcinoma of the bladder (prostate)	49	358
Metastatic tumours (lymph node)	50	372–379
Papilloma (breast)	46	322–324
Prostatitis	48	344, 345

	Page	Figures

FINE NEEDLE ASPIRATION CYTOLOGY (Continued)

	Page	Figures
Reticulosarcoma (lymph node)	50	365, 366
Sarcoma (breast)	48	319–321
Thyroiditis	44, 45	278–280
acute	44	—
lymphoid, chronic	45	279, 280
subacute	43	278

Lung

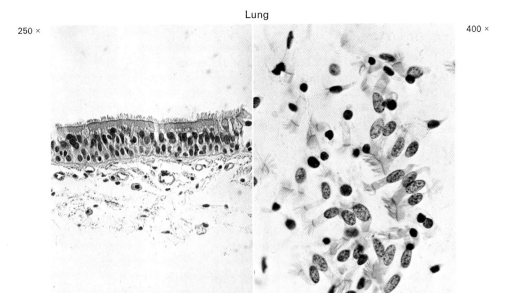

Fig. 1 — Bronchial epithelium
Fig. 2 — Normal bronchial epithelial cells

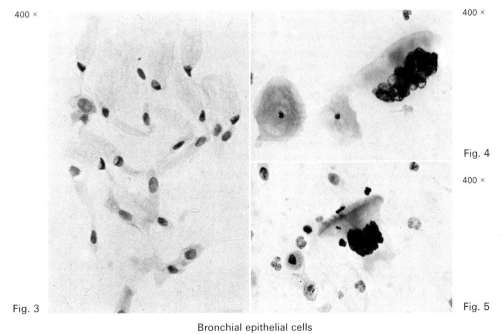

Fig. 3 — Goblet cells
Fig. 4, Fig. 5 — Bronchial epithelial cells, Multinucleation

Lung

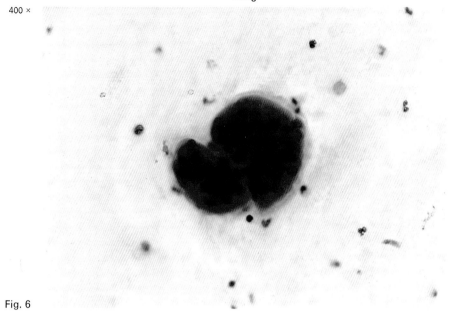

Fig. 6

Bronchial epithelial cell cluster

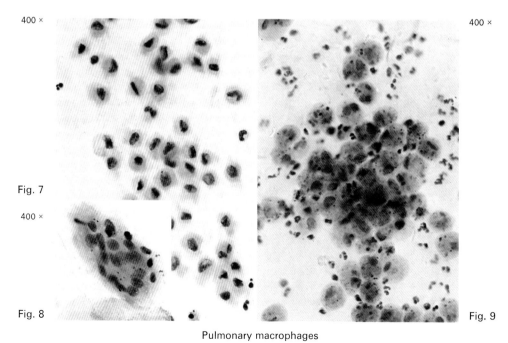

Fig. 7

Fig. 8   Fig. 9

Pulmonary macrophages

Non-pigmented   Pigmented

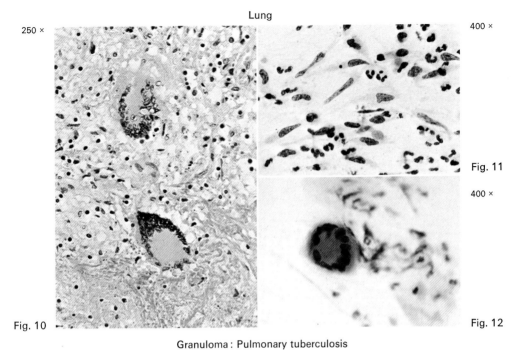

Lung

250 ×

Fig. 10

400 ×

Fig. 11

400 ×

Fig. 12

Granuloma: Pulmonary tuberculosis

Histologic section

Fig. 11. Epithelioid cells in sputum
Fig. 12. Langhans giant cell in sputum

400 ×

Fig. 13

400 ×

Fig. 14

400 ×

Fig. 15

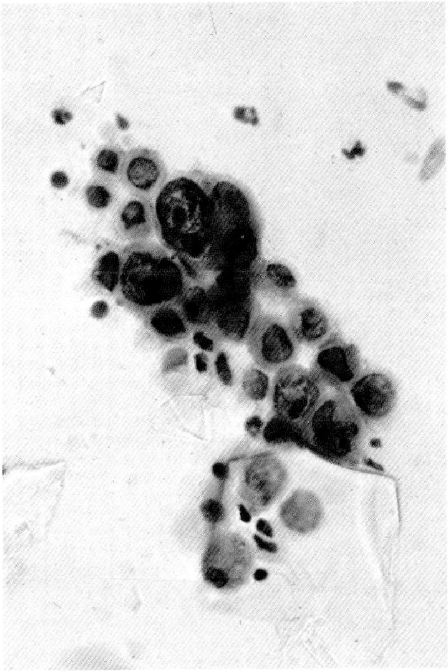

Viral changes

Clusters of epithelial cells

Multinucleated giant cell

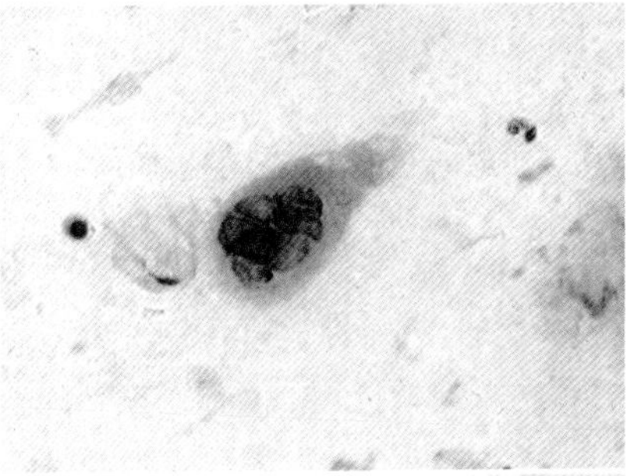

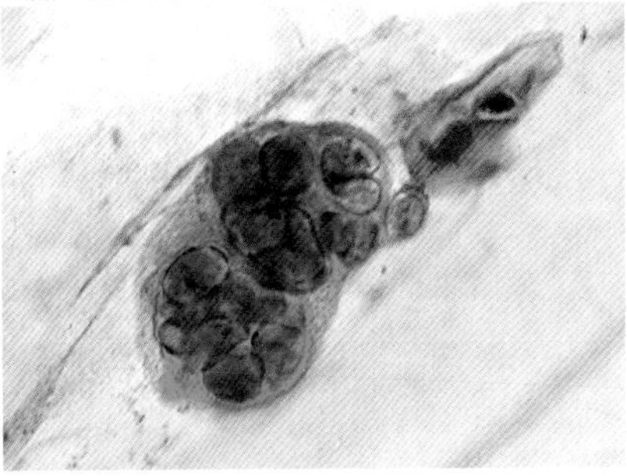

Lung

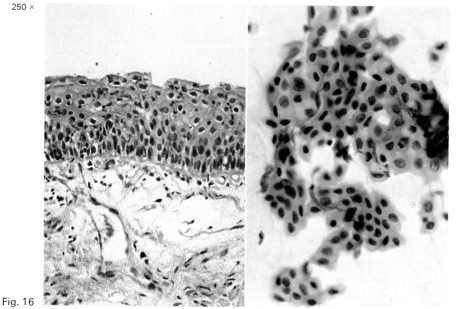

Fig. 16 — 250×
Fig. 17 — 400×

Squamous metaplasia

Bronchial epithelium   Bronchial aspiration

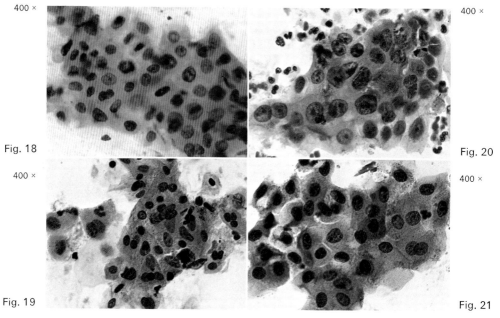

Fig. 18 — 400×
Fig. 19 — 400×
Fig. 20 — 400×
Fig. 21 — 400×

Atypical squamous metaplasia: sputum
Mild atypia

Lung

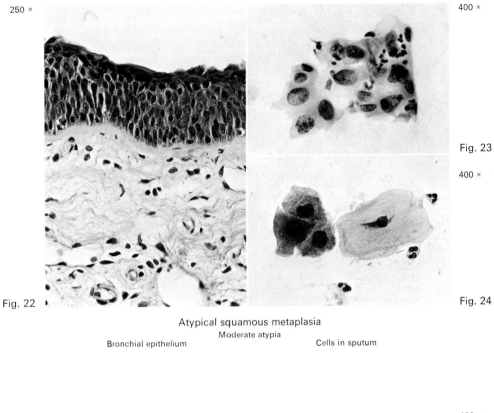

Atypical squamous metaplasia
Moderate atypia
Bronchial epithelium   Cells in sputum

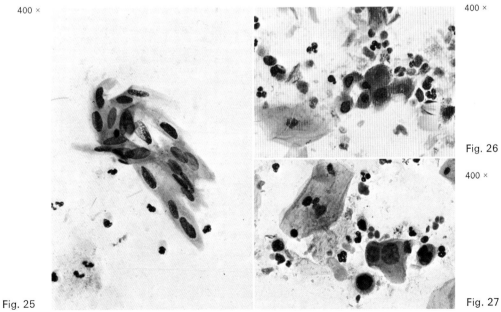

Atypical squamous metaplasia: sputum
Severe atypia

Lung

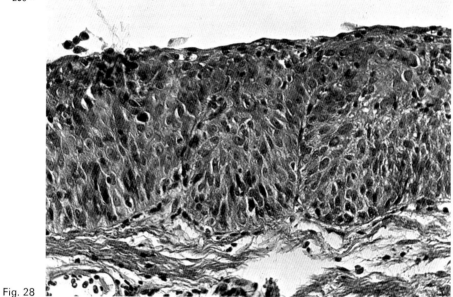

Fig. 28
Squamous cell carcinoma in situ

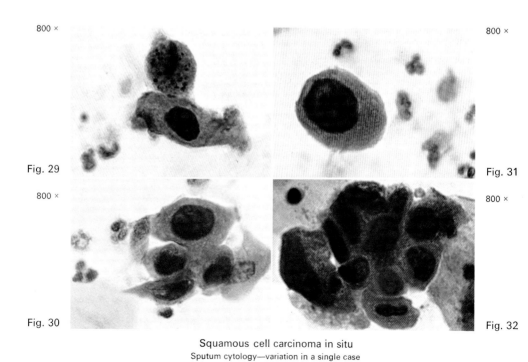

Fig. 29
Fig. 30
Fig. 31
Fig. 32

Squamous cell carcinoma in situ
Sputum cytology—variation in a single case

Lung

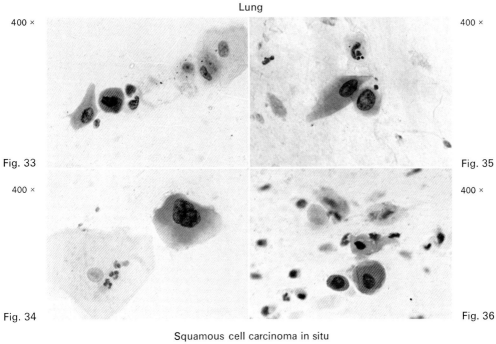

Fig. 33
Fig. 35
Fig. 34
Fig. 36

Squamous cell carcinoma in situ
Sputum cytology—variation in a single case

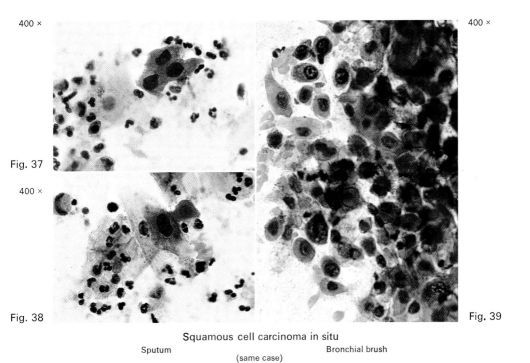

Fig. 37
Fig. 38
Fig. 39

Squamous cell carcinoma in situ

Sputum   Bronchial brush
(same case)

Lung

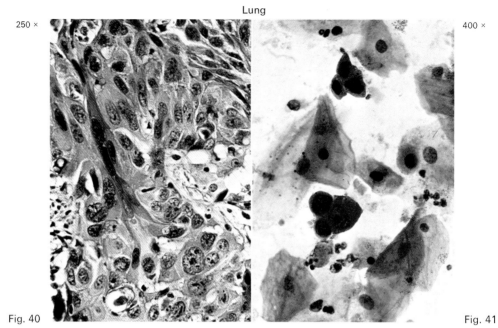

Fig. 40 — 250 ×
Fig. 41 — 400 ×

Invasive squamous cell carcinoma: Moderate differentiation
Tissue   Sputum

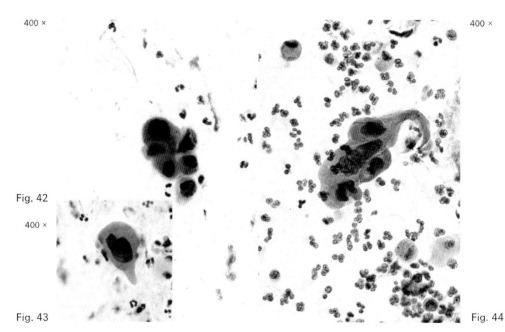

Fig. 42 — 400 ×
Fig. 43 — 400 ×
Fig. 44 — 400 ×

Invasive squamous cell carcinoma
Clumps and single carcinoma cell: sputum

Lung

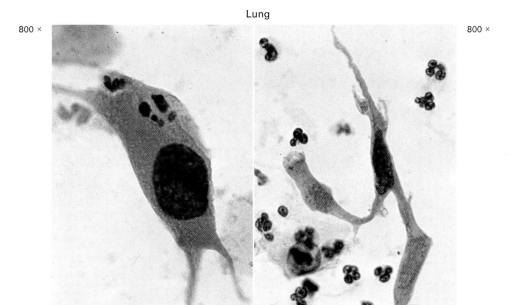

Fig. 45   Fig. 46

Invasive squamous cell carcinoma: sputum
Bizarre cell   Spindle cell

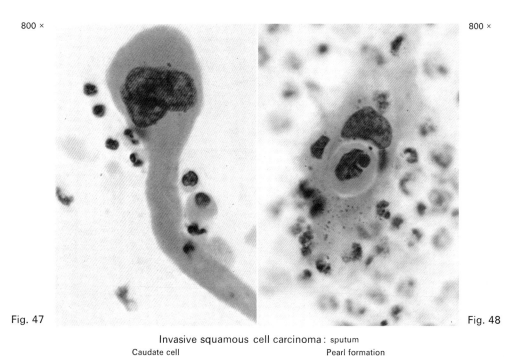

Fig. 47   Fig. 48

Invasive squamous cell carcinoma: sputum
Caudate cell   Pearl formation

Lung

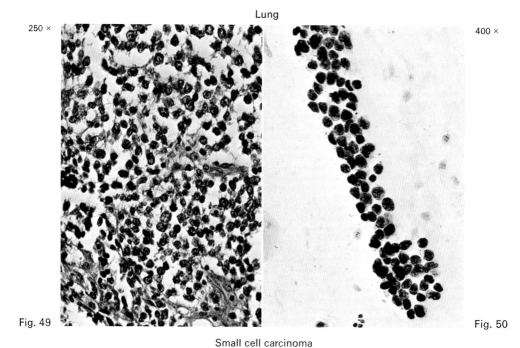

Small cell carcinoma

Tissue section   Sputum (same case)

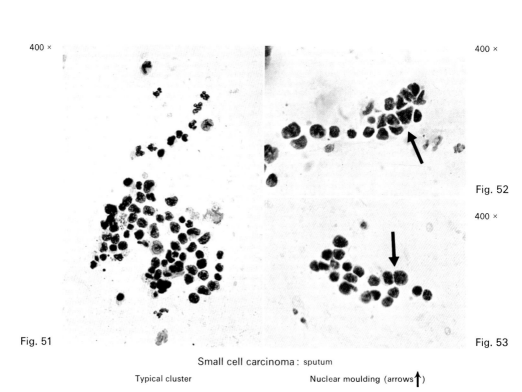

Small cell carcinoma : sputum

Typical cluster   Nuclear moulding (arrows↑)

Lung

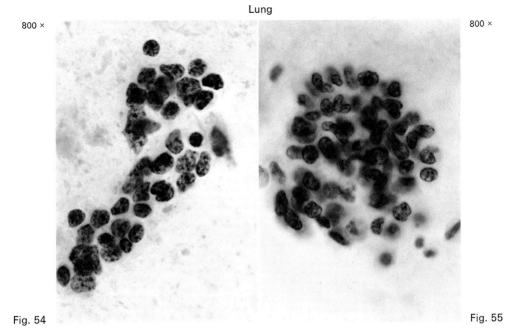

Fig. 54 — Small cell carcinoma
Cluster of carcinoma cells

Fig. 55 — Small histiocytes, for comparison

Sputum

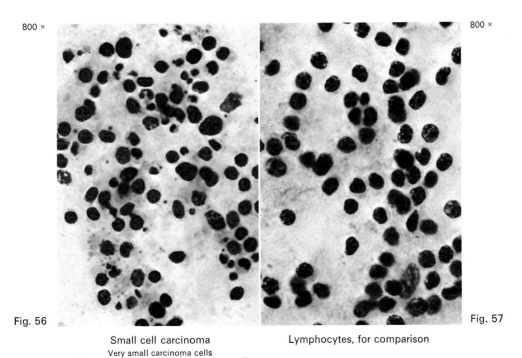

Fig. 56 — Small cell carcinoma
Very small carcinoma cells

Fig. 57 — Lymphocytes, for comparison

Sputum

Lung

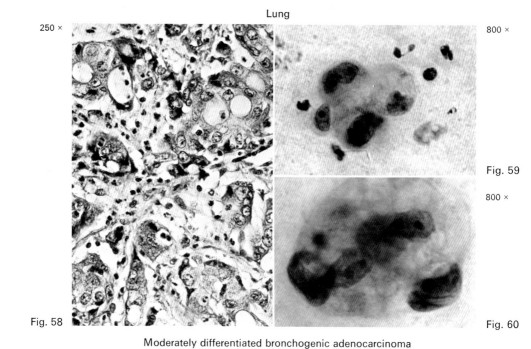

Moderately differentiated bronchogenic adenocarcinoma
Histology   Sputum (same case)

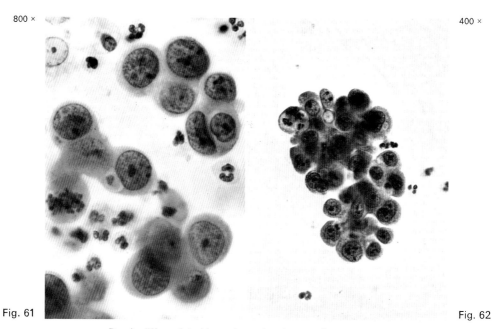

Poorly differentiated bronchogenic adenocarcinoma: sputum

Lung

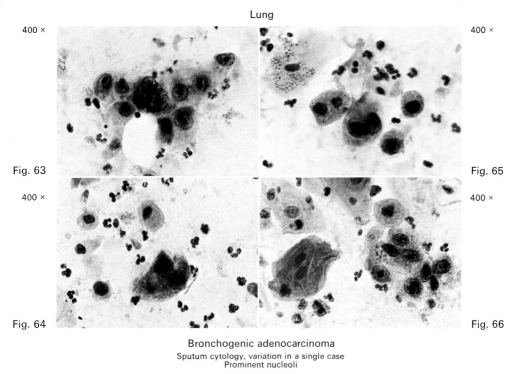

Fig. 63
Fig. 65
Fig. 64
Fig. 66

Bronchogenic adenocarcinoma
Sputum cytology, variation in a single case
Prominent nucleoli

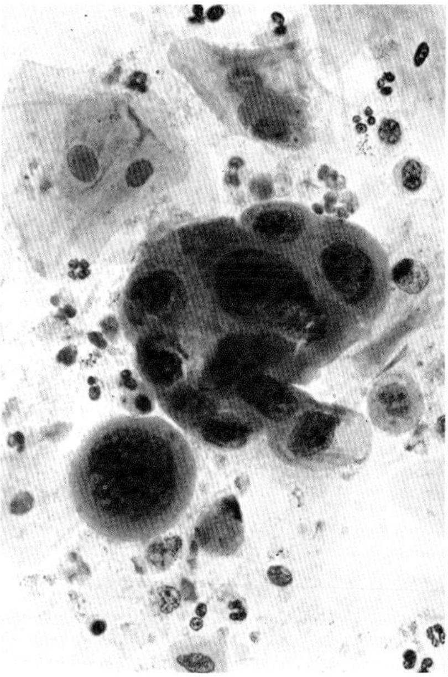

Fig. 67
Fig. 68

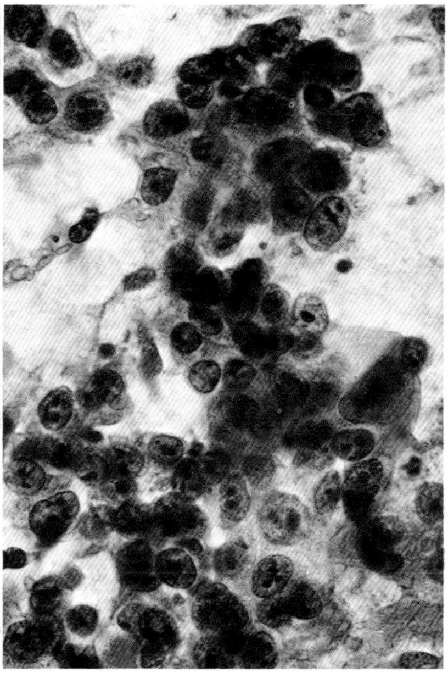

Poorly differentiated bronchogenic adenocarcinoma
Sputum   Brush cytology (same case)

Lung

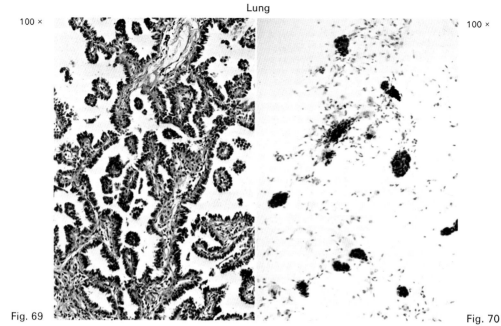

Fig. 69 — 100×
Fig. 70 — 100×

Well differentiated bronchiolo-alveolar adenocarcinoma
Histology | Bronchial aspiration, numerous cell clumps (same case)

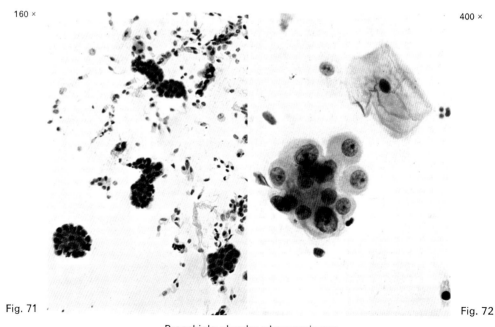

Fig. 71 — 160×
Fig. 72 — 400×

Bronchiolo-alveolar adenocarcinoma
Bronchial aspiration, higher magnification, same case as Fig. 70 | Typical cell clumps: sputum

Lung

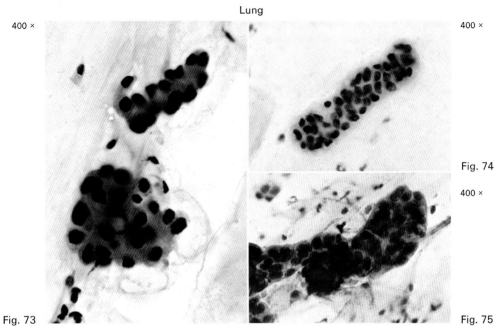

Bronchiolo-alveolar adenocarcinoma: papillary cell clumps
Bronchial aspiration   Sputum

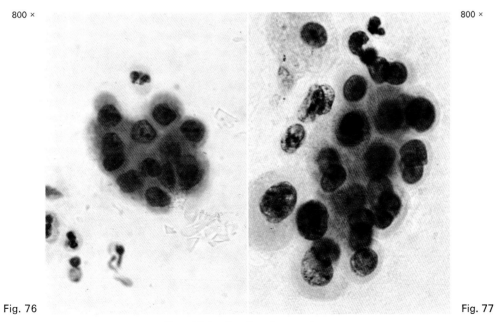

Bronchiolo-alveolar adenocarcinoma
Typical cell clumps: sputum

Lung

250 × 400 ×

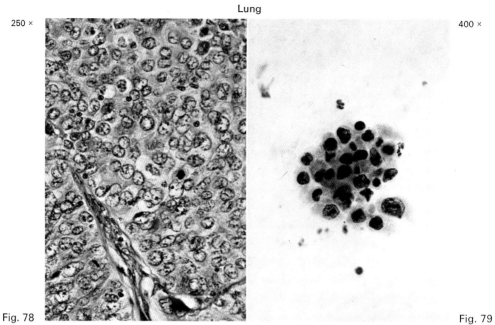

Fig. 78 Fig. 79

Large cell carcinoma
Histology Sputum (same case)

800 ×

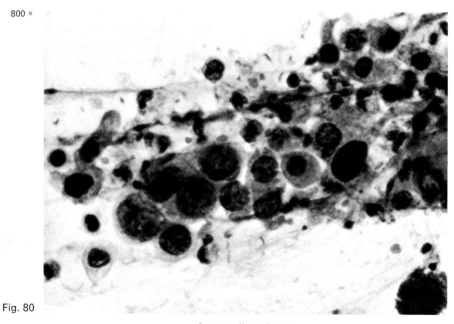

Fig. 80

Large cell carcinoma
Undifferentiated carcinoma cells: sputum

Lung

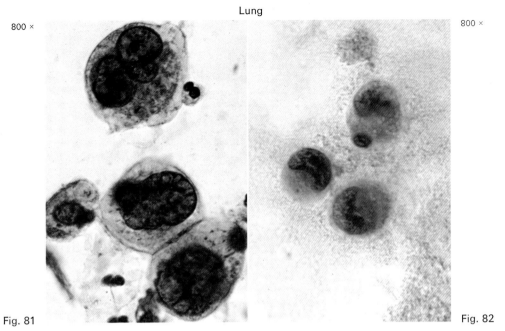

Fig. 81 Fig. 82

Large cell carcinoma
Undifferentiated carcinoma cells: sputum

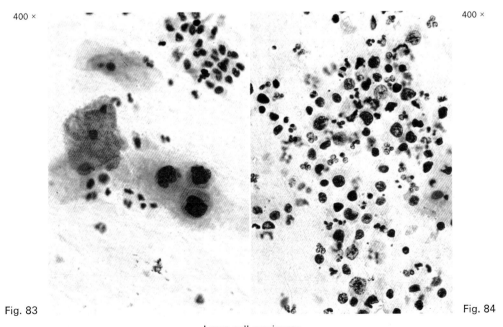

Fig. 83 Fig. 84

Large cell carcinoma
Undifferentiated carcinoma cells: sputum

Lung

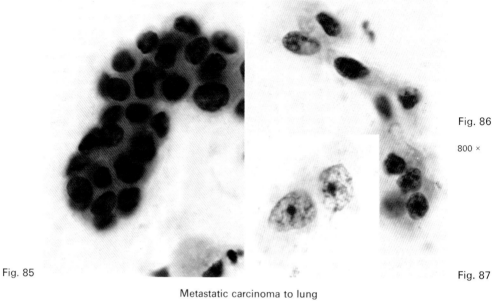

Fig. 85 — Mammary carcinoma
Fig. 86 — Kidney: clear cell carcinoma
Fig. 87

Metastatic carcinoma to lung
Sputum

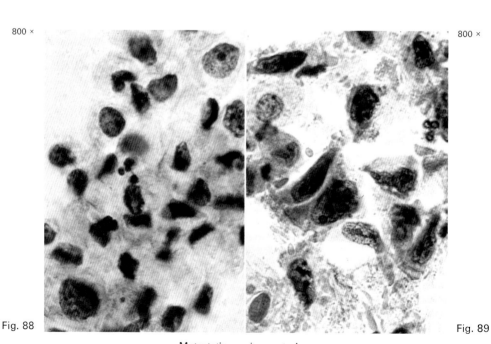

Fig. 88 — Gastric carcinoma
Fig. 89 — Colon carcinoma

Metastatic carcinoma to lung
Sputum

Lung

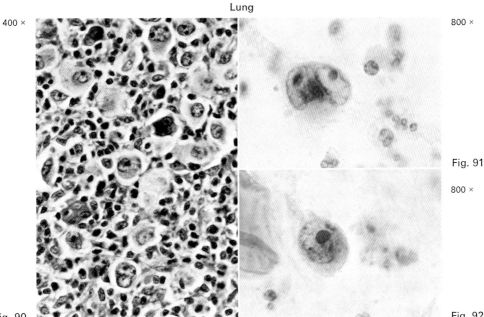

400 ×
800 ×
Fig. 91
800 ×
Fig. 90
Fig. 92

Hodgkin's disease

Histology
Fig. 91. Reed-Sternberg cells: sputum
Fig. 92. Mononucleated cell: sputum
(same case)

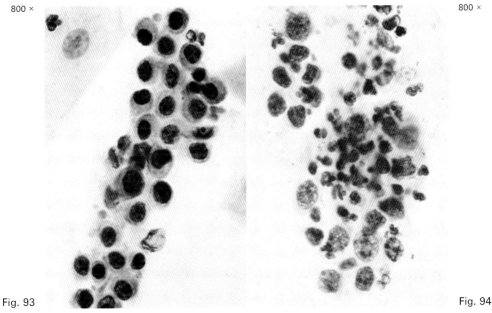

800 ×
800 ×
Fig. 93
Fig. 94

Plasmocytoma: sputum
Reticulosarcoma: sputum

Lung

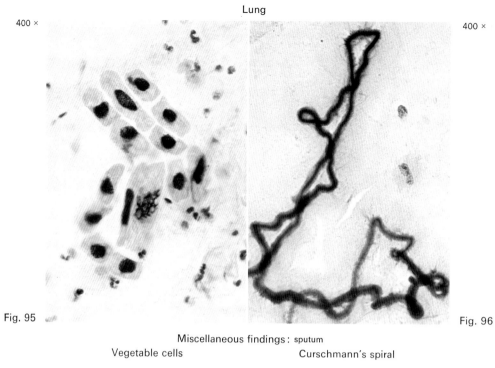

Fig. 95　　　　　　　　　　　　　　　　　　　　　　　　　　　　　Fig. 96

Miscellaneous findings: sputum

Vegetable cells　　　　　　　　　　Curschmann's spiral

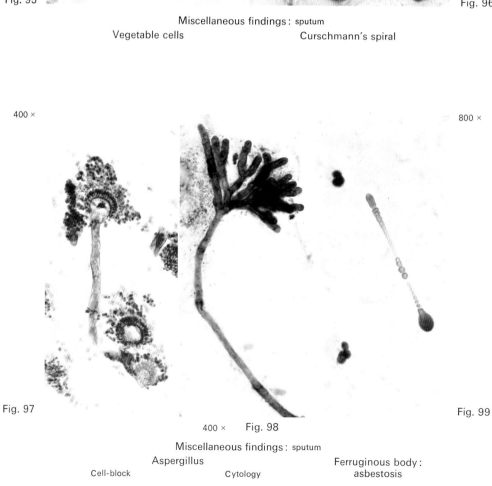

Fig. 97　　　　　　　　　　　　　　　　　　　　　　　　　　　　　Fig. 99

400 ×　　Fig. 98

Miscellaneous findings: sputum

　　　　　　Aspergillus　　　　　　　　　　Ferruginous body:
Cell-block　　　　　　Cytology　　　　　　　　asbestosis

Urinary tract

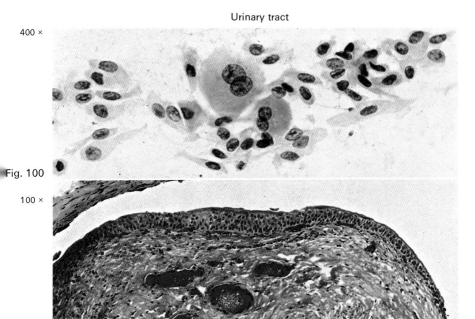

Fig. 100 400 ×
Fig. 101 100 ×

Bladder: Normal urothelium

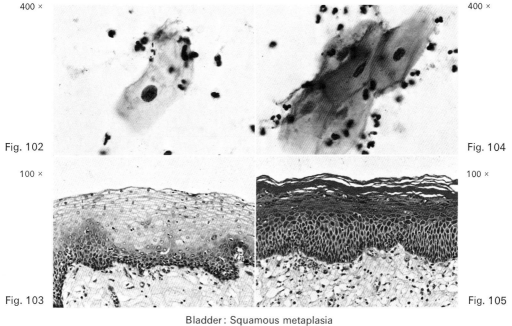

Fig. 102 400 ×
Fig. 104 400 ×
Fig. 103 100 ×
Fig. 105 100 ×

Bladder: Squamous metaplasia
Non-keratinizing   Keratinizing

Urinary tract

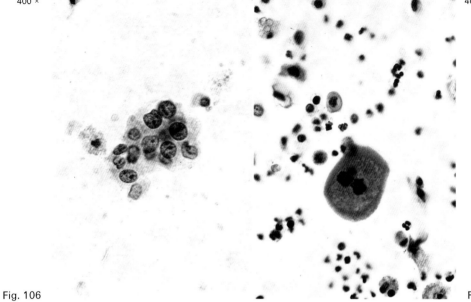

Fig. 106
Epithelial cells in acute cystitis

Fig. 107
Epithelial cells in chronic cystitis

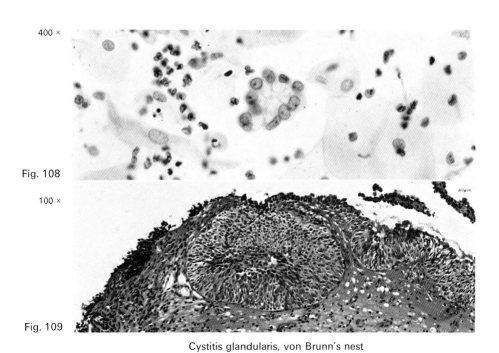

Fig. 108

Fig. 109
Cystitis glandularis, von Brunn's nest

Urinary tract

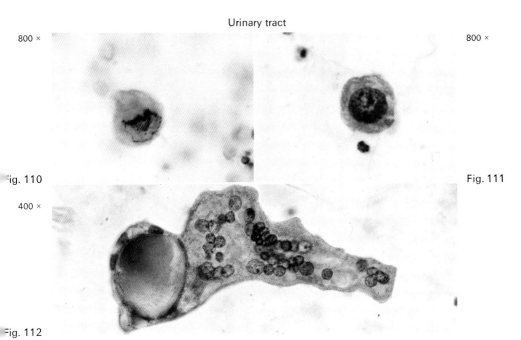

Fig. 110. Renal tubular epithelial cell   Fig. 111. Cell from seminal vesicles
Fig. 112. Cells avulsed by ureteric catheter

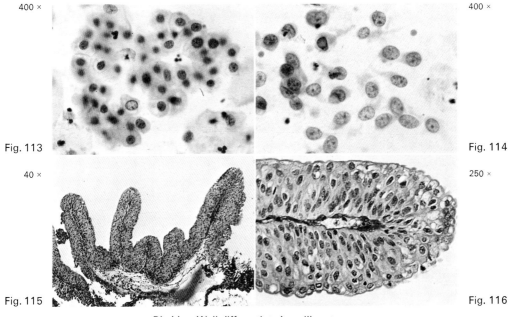

Bladder: Well differentiated papillary tumour
Fig 113. Papillary clump—cytology—Fig 114. Individual cells
Fig. 115/116. Histology

Urinary tract

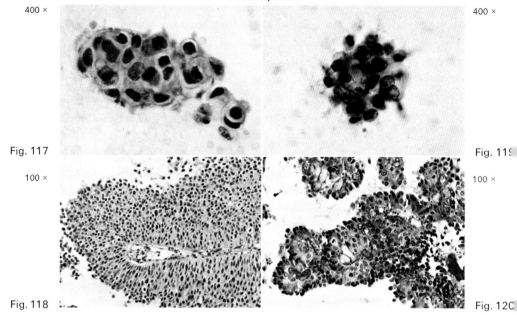

Fig. 117 400 ×
Fig. 119 400 ×
Fig. 118 100 ×
Fig. 120 100 ×

Bladder: Papillary carcinoma
Moderately differentiated   Poorly differentiated

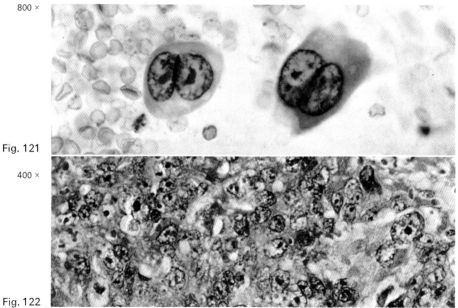

Fig. 121 800 ×
Fig. 122 400 ×

Bladder: Non-papillary carcinoma

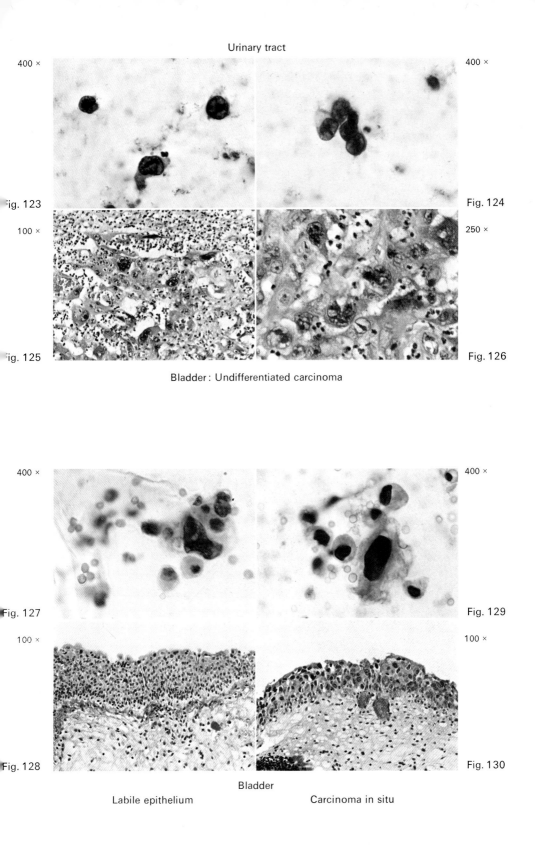

Urinary tract

Fig. 123 400×
Fig. 124 400×
Fig. 125 100×
Fig. 126 250×

Bladder: Undifferentiated carcinoma

Fig. 127 400×
Fig. 129 400×
Fig. 128 100×
Fig. 130 100×

Bladder

Labile epithelium   Carcinoma in situ

Urinary tract

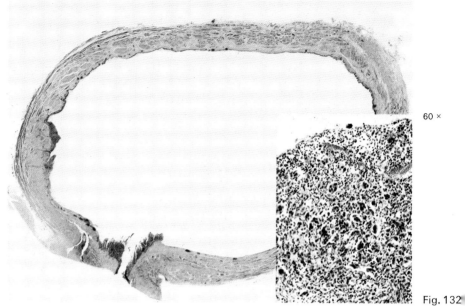

Fig. 131   1×
Fig. 132   60×

Total bladder section: carcinoma in situ with zone of invasion
Insert: invasive carcinoma

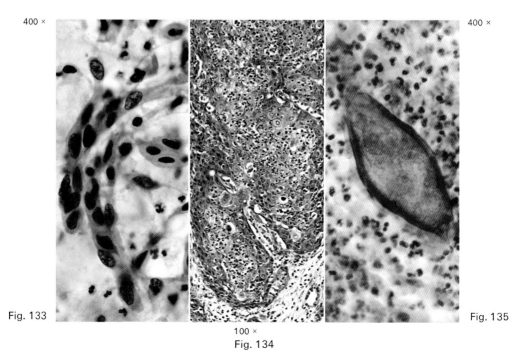

Fig. 133   400×
Fig. 134   100×
Fig. 135   400×

Bladder: Squamous cell carcinoma   Schistosomiasis
                                          (Schistosoma hematobium)

Urinary tract

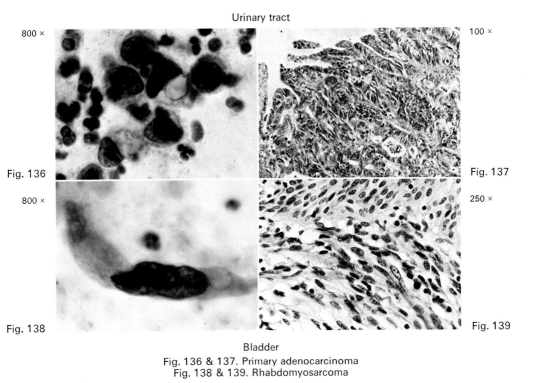

Bladder
Fig. 136 & 137. Primary adenocarcinoma
Fig. 138 & 139. Rhabdomyosarcoma

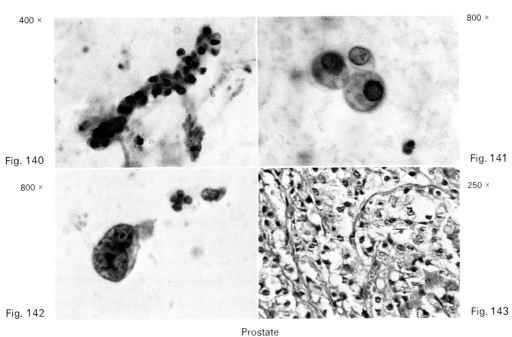

Prostate
Fig. 140. Normal duct epithelium
Fig. 142. Poorly differentiated carcinoma
Fig. 141. Well differentiated carcinoma
Fig. 143. Histology of prostatic carcinoma

Urinary tract

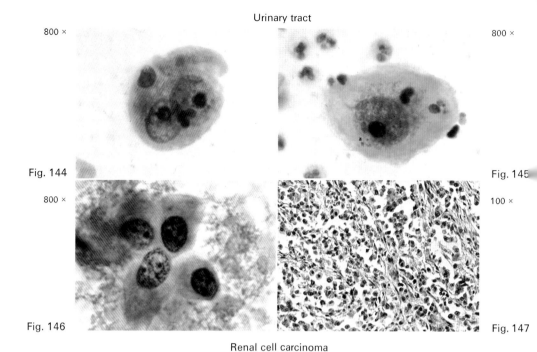

Renal cell carcinoma

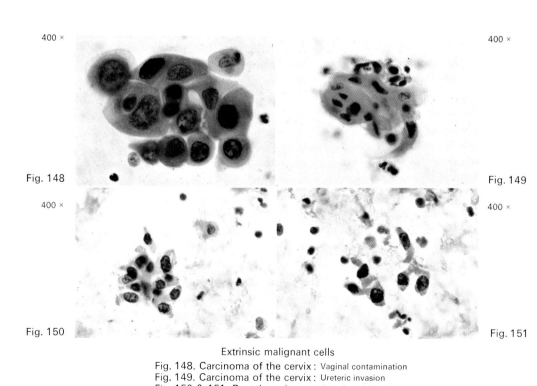

Extrinsic malignant cells
Fig. 148. Carcinoma of the cervix: Vaginal contamination
Fig. 149. Carcinoma of the cervix: Ureteric invasion
Fig. 150 & 151. Rectal carcinoma: Invasion of bladder

Urinary tract

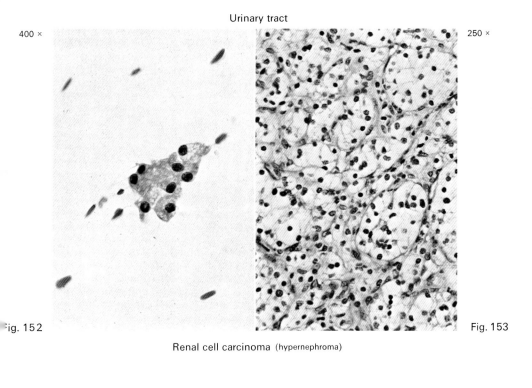

Fig. 152
Fig. 153
Renal cell carcinoma (hypernephroma)

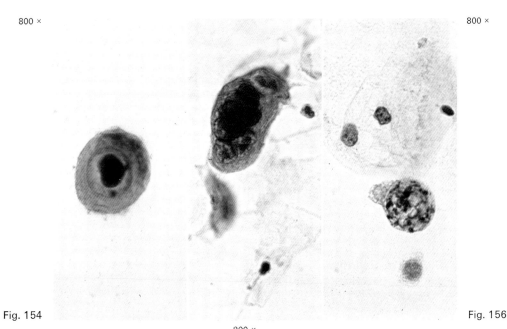

Fig. 154
Fig. 155
Fig. 156
Urine: miscellaneous findings
Cytomegalic inclusion Herpes Decoy cell

Body fluids

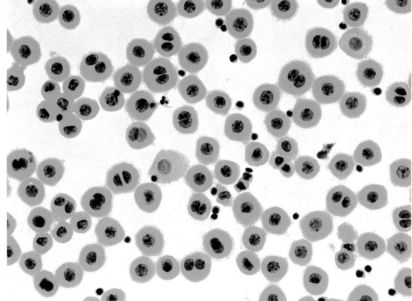

Fig. 157

Normal mesothelial cells

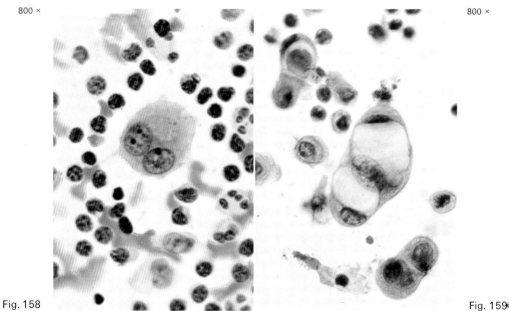

Fig. 158

Normal mesothelial cells, lymphocytes, histiocytes and polymorphonuclear leukocytes

Fig. 159

Vacuolated mesothelial cells + cytophagia and engulfment

Body fluids

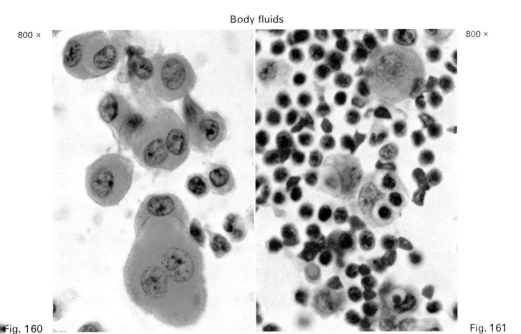

Fig. 160 — Cirrhosis of liver
Note enlarged binucleated mesothelial cells

Fig. 161 — Tuberculous peritonitis
Note phagocytosis of lymphocytes

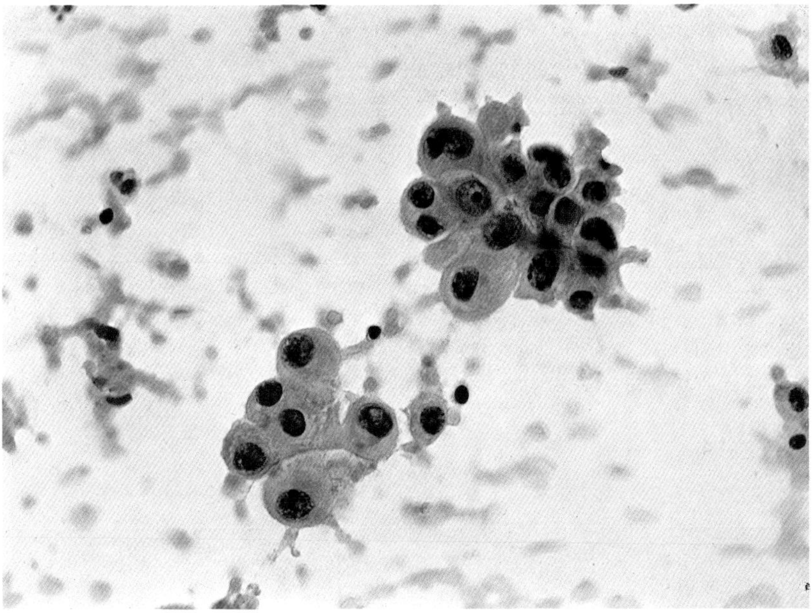

Fig. 162 — Mesothelioma
The sharp borders with clear areas between cells are characteristic
for cells of mesothelial origin

Body fluids

800 × 800 ×

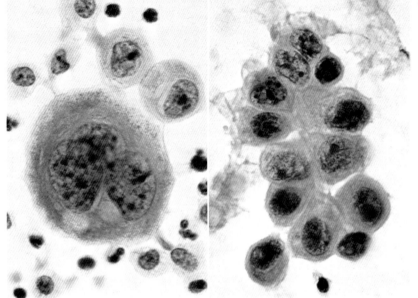

Fig. 163　　　　　　　　　　　　　　　　　　　　　　　　　　　Fig. 164

Mesothelioma

160 ×

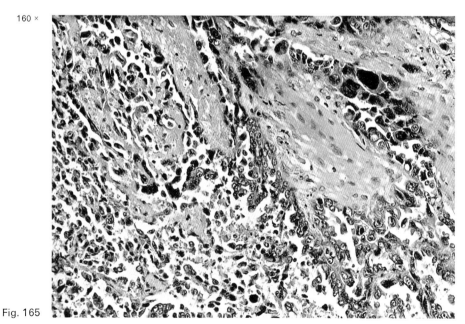

Fig. 165

Mesothelioma

Body fluids

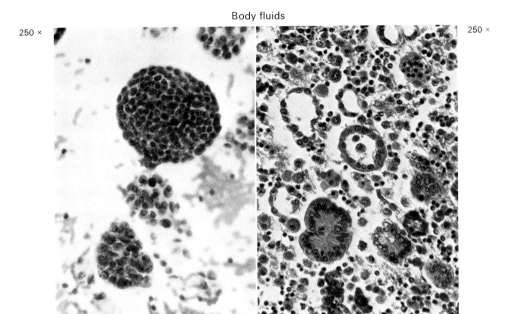

Fig. 166   Fig. 167

Breast metastasis to pleura

Cytology   Cell-block (same case)

The "ball formation" is quite characteristic of metastatic breast cancer.
The cell-block shows the appearance of the "balls" on sectioning

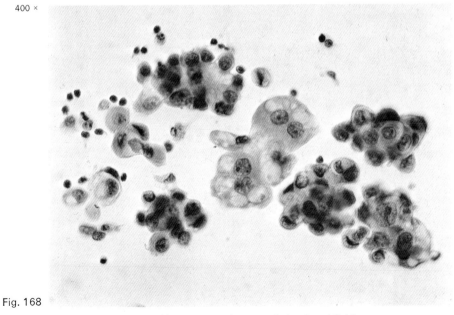

Fig. 168

Mammary carcinoma cells in pleural fluid

Body fluids

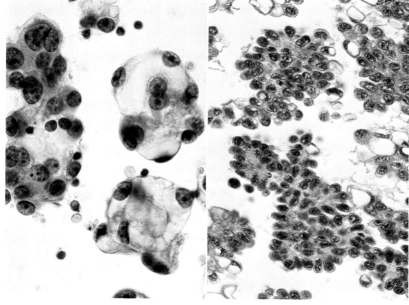

Fig. 169   Fig. 170

Metastatic serous cystadenocarcinoma of ovary: Ascitic fluid
Cytology   Cell-block (same case)

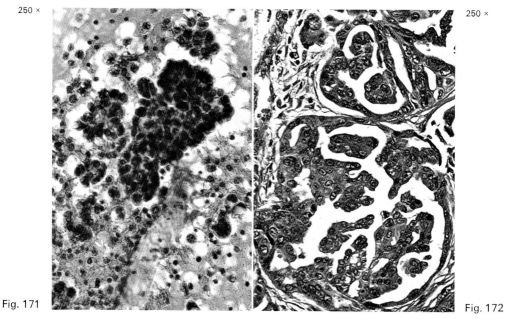

Fig. 171   Fig. 172

Metastatic endometrioid adenocarcinoma of ovary: Ascitic fluid
Cytology   Primary tumour histology (same case)

Body fluids

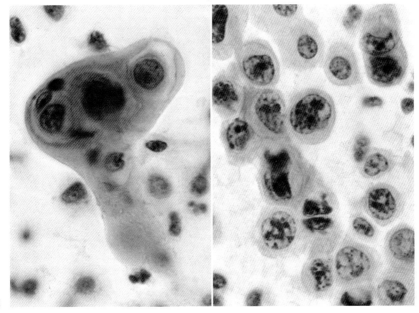

800 ×   Fig. 173

800 ×   Fig. 174

Metastatic squamous cell carcinoma of cervix: well differentiated   Metastatic squamous cell carcinoma of cervix: poorly differentiated

Ascitic fluid

800 ×   Fig. 175

250 ×   Fig. 176

Metastatic oat cell carcinoma of lung

Pleural fluid   Primary tumour histology (same case)
Note nuclear moulding (arrows ↑)

Body fluids

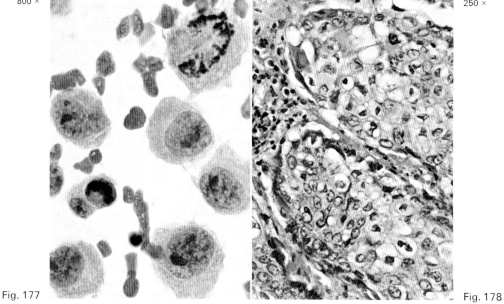

Fig. 177 — 800 ×
Fig. 178 — 250 ×

Metastatic large cell carcinoma of lung
Pleural fluid   Primary tumour histology (same case)

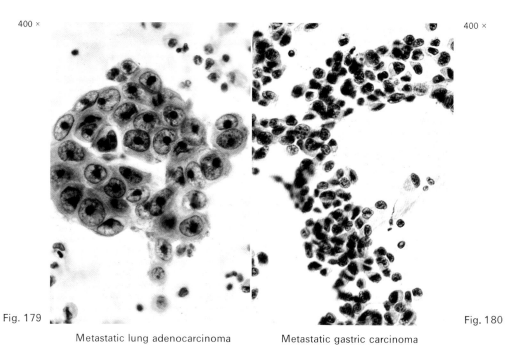

Fig. 179 — 400 ×
Fig. 180 — 400 ×

Metastatic lung adenocarcinoma
Pleural fluid

Metastatic gastric carcinoma
poorly differentiated
Ascitic fluid

Body fluids

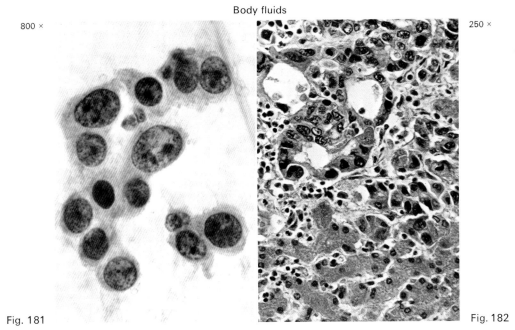

Fig. 181 — Ascitic fluid
Fig. 182 — Histology (same case)

Primary liver carcinoma: bile duct type

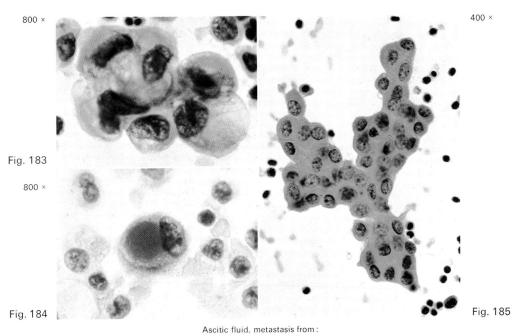

Ascitic fluid, metastasis from:

Fig. 183. Pancreatic carcinoma
Fig. 184. Pancreatic carcinoma (PAS)
Fig. 185. Adenocarcinoma from caecum

Body fluids

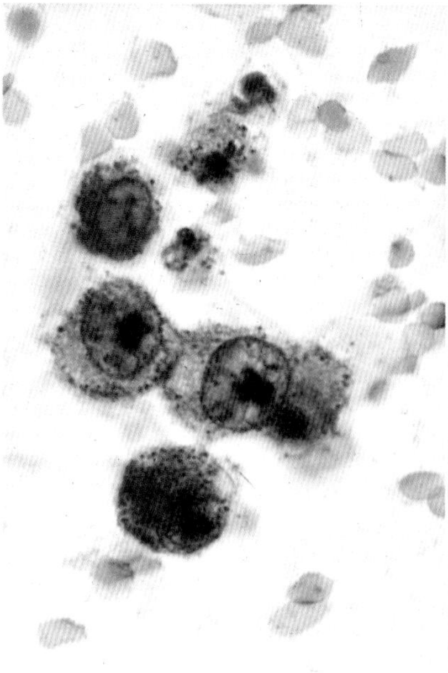

Melanoma

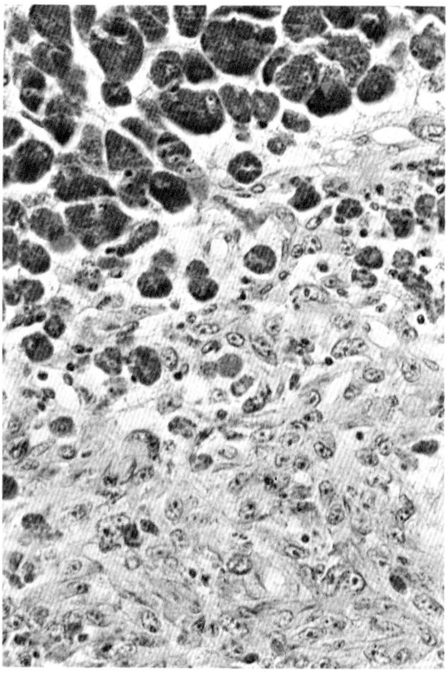

Fig. 186 — Metastasis, ascitic fluid (800 ×)
Fig. 187 — Histology (same case) (250 ×)

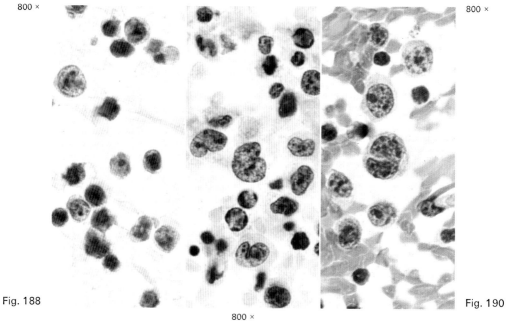

Fig. 188 — Lymphosarcoma: lymphocytic type (800 ×)
Fig. 189 — Reticulosarcoma, Ascitic fluid (800 ×)
Fig. 190 — Hodgkin's disease (800 ×)

Oesophagus

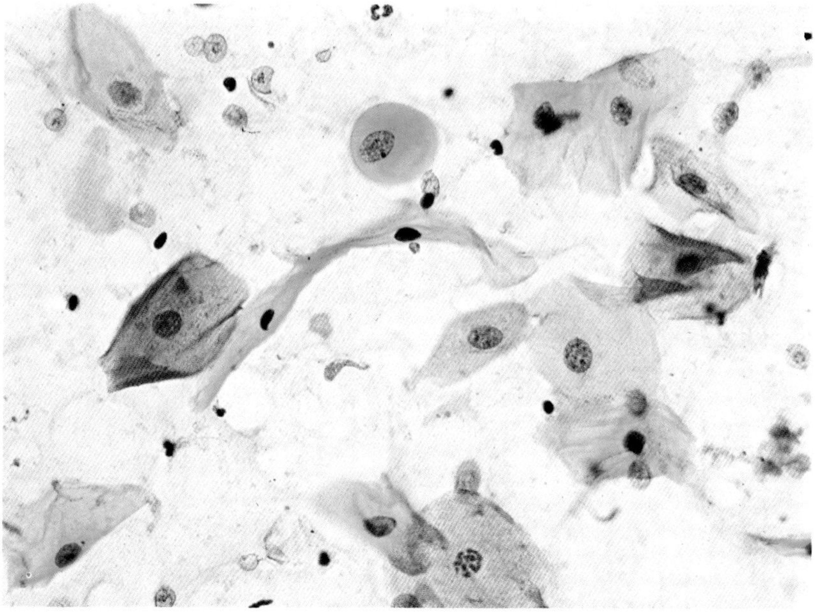

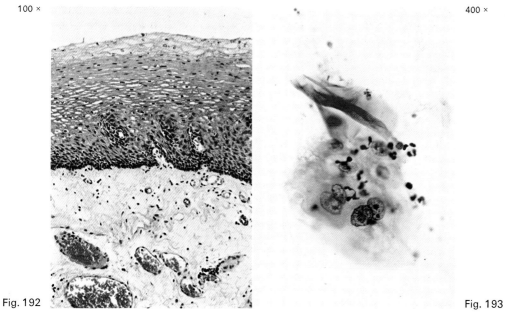

Fig. 191

Normal squamous epithelium from oesophagus: Brush

Fig. 192

Section from normal squamous epithelium from oesophagus

Fig. 193

Oesophagitis
Inflammatory changes and keratinization in squamous cells: Brush

Oesophagus

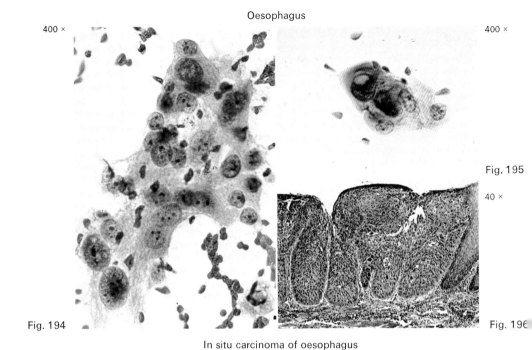

In situ carcinoma of oesophagus
Note that in brush specimen the cell appearance can be typically squamous or,
as syncytial sheets, reminiscent of a glandular type of lesion

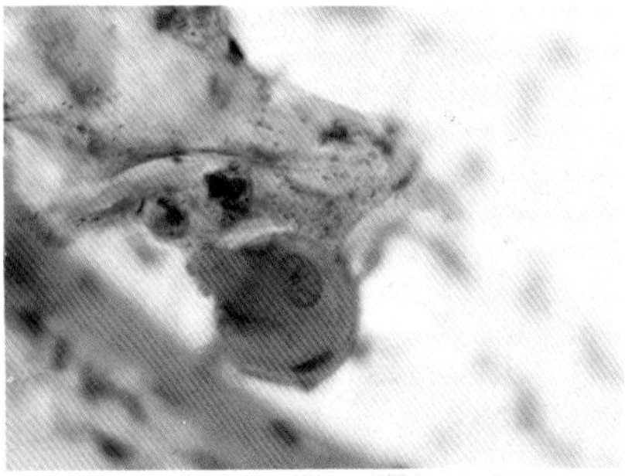

Epithelial pearls
Fig. 197. Benign: oesophagitis
Fig. 198. Malignant: well differentiated squamous cell carcinoma

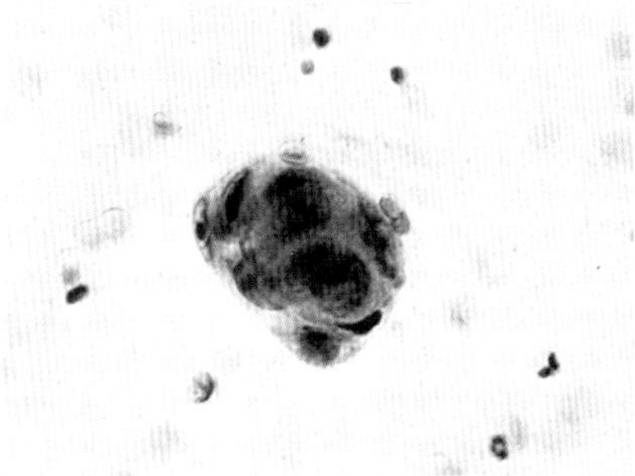

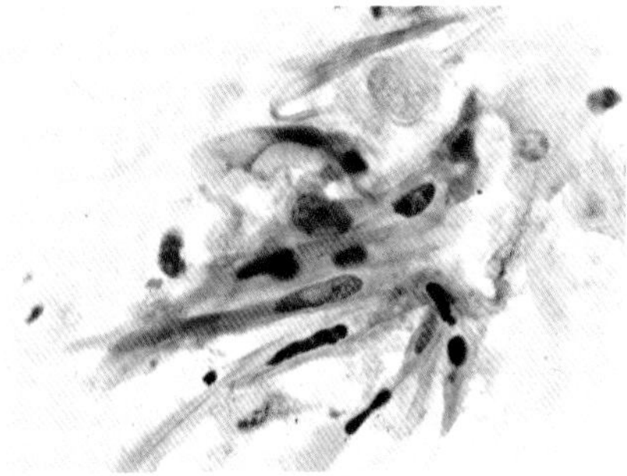

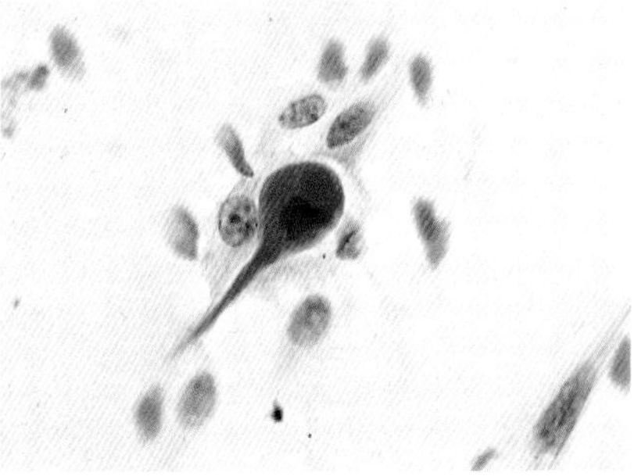

Squamous cell carcinoma of oesophagus
Fig. 199. Fibre type cells
Fig. 200. Caudate cell

Stomach

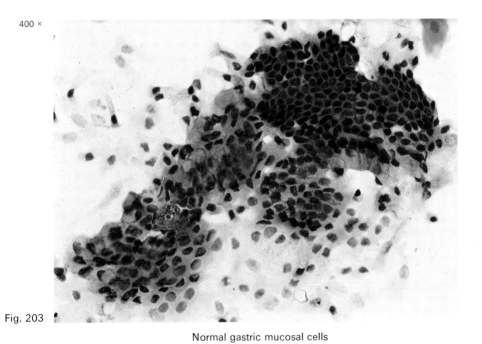

60 ×   60 ×

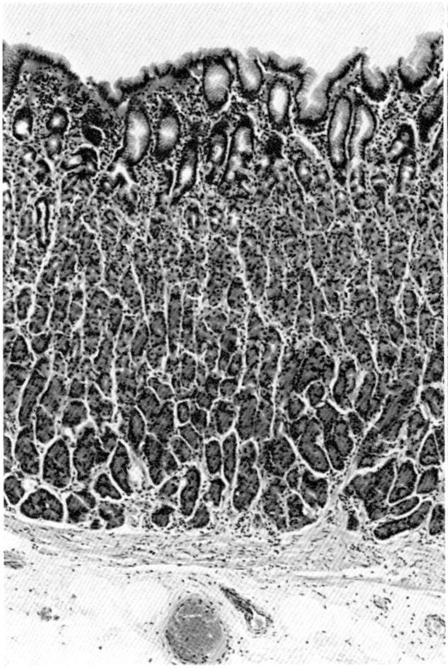

Fig. 201   Fig. 202

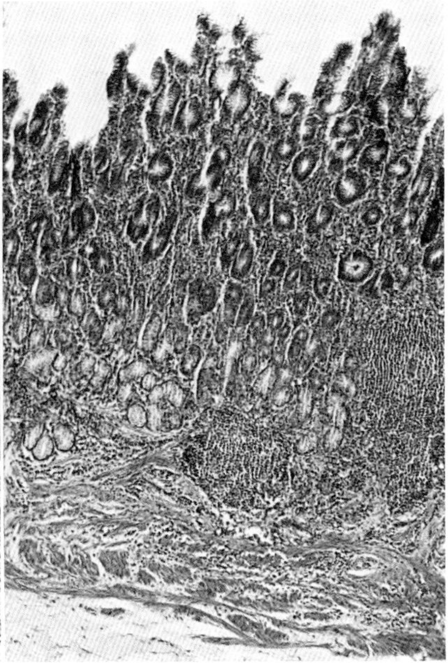

Normal gastric mucosa

Fundus   Pyloris

400 ×

Fig. 203

Normal gastric mucosal cells

Stomach

250 ×

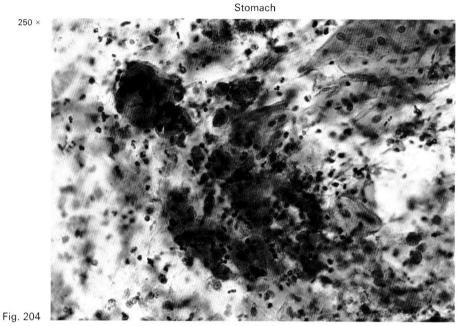

Fig. 204

Normal gastric mucosal cells
Mild inflammatory changes

400 ×   400 ×

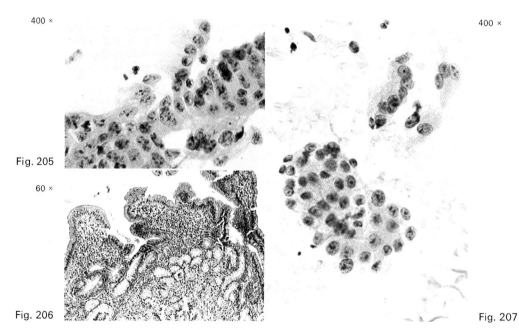

Fig. 205

60 ×

Fig. 206   Fig. 207

Atrophic gastritis in a case of gastric ulcer

Stomach

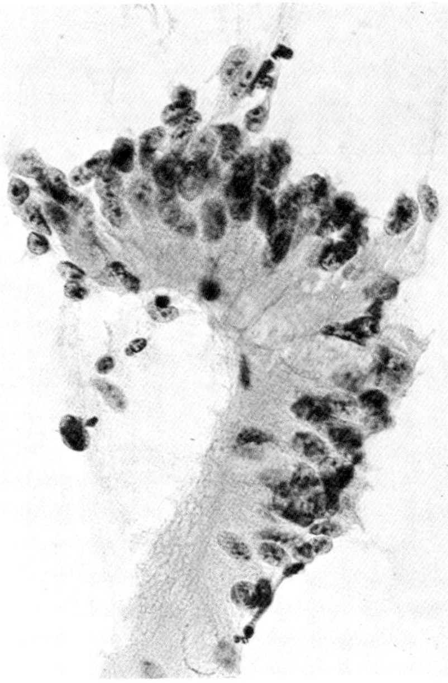

400 × 60 ×

Fig. 208 Fig. 209

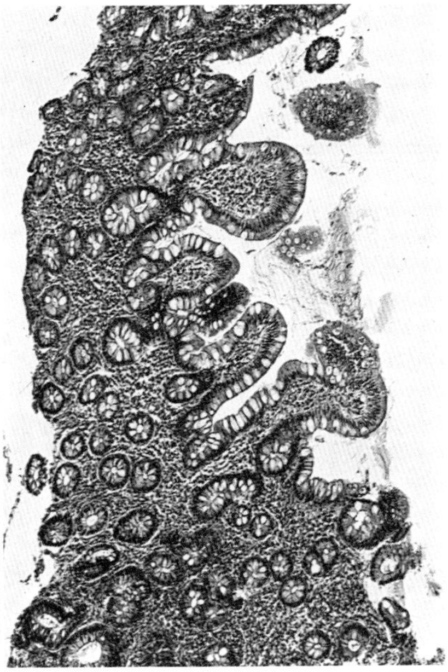

Atrophic gastritis
Note inflammatory and degenerative changes in an atrophic epithelium

400 × 60 ×

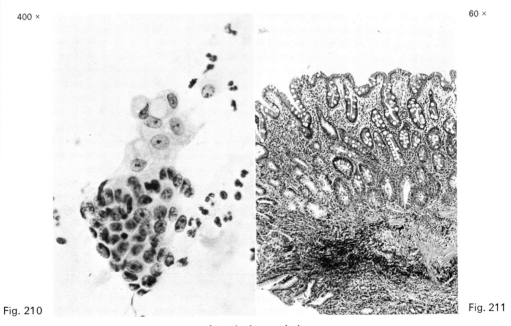

Fig. 210 Fig. 211

Intestinal metaplasia
Note vesicular type of goblet cell formation

Stomach

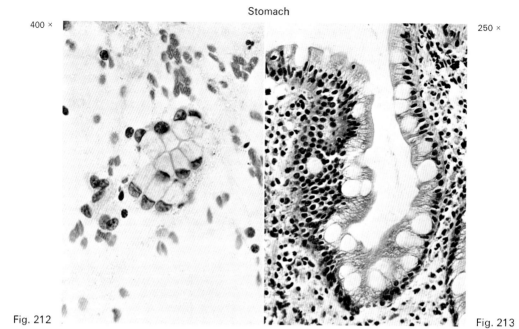

400 ×　Fig. 212　　250 ×　Fig. 213

Mild intestinal metaplasia with goblet cell formation

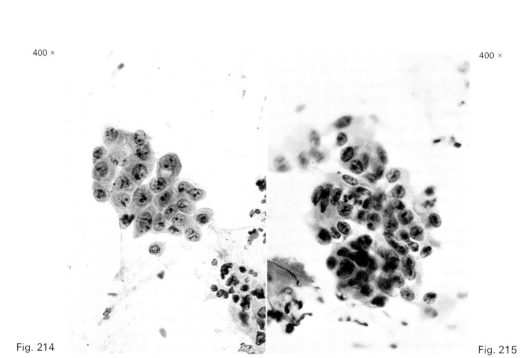

400 ×　Fig. 214　　400 ×　Fig. 215

Regeneration changes in gastric ulcer

Stomach

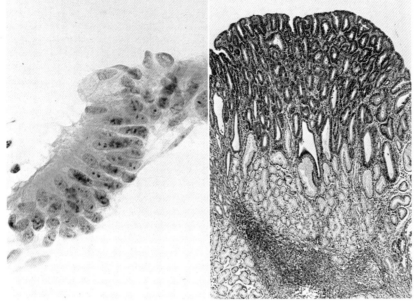

Fig. 216 Fig. 217

Atypical epithelium in metaplasia
Note the fusiform appearance of the cells and nuclei

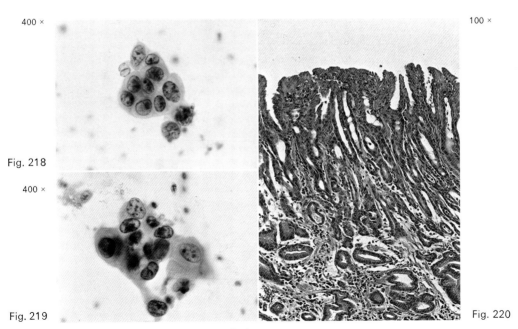

Fig. 218
Fig. 219 Fig. 220

Surface cancer

Stomach

400 ×
2.5 ×

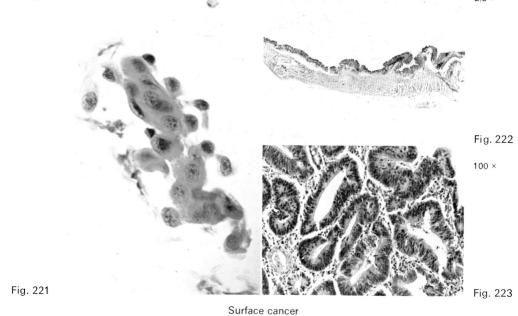

Fig. 222

100 ×

Fig. 221
Fig. 223

Surface cancer

400 ×
2.5 ×

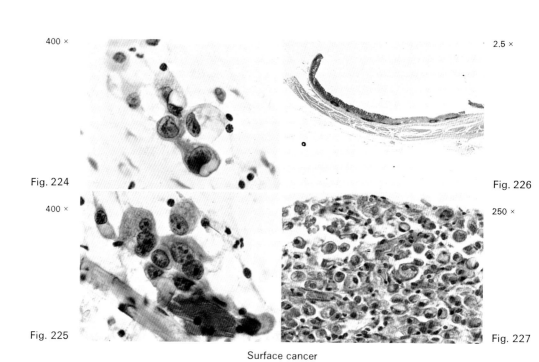

Fig. 224
Fig. 226

400 ×
250 ×

Fig. 225
Fig. 227

Surface cancer

Stomach

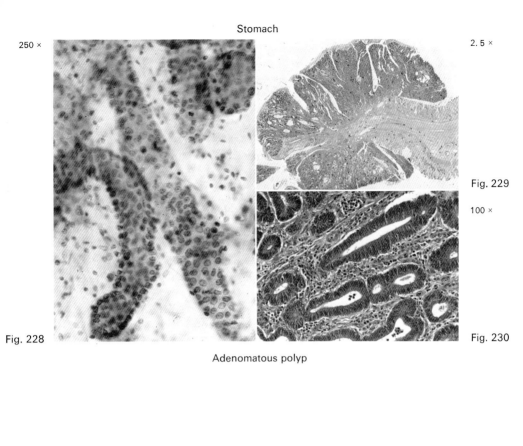

Fig. 228 250×
Fig. 229 2.5×
Fig. 230 100×

Adenomatous polyp

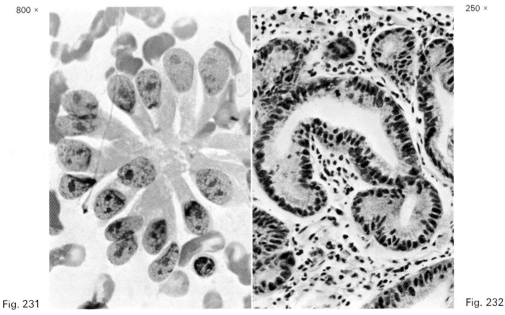

Fig. 231 800×
Fig. 232 250×

Adenomatous polyp
Cytology: markedly active

Stomach

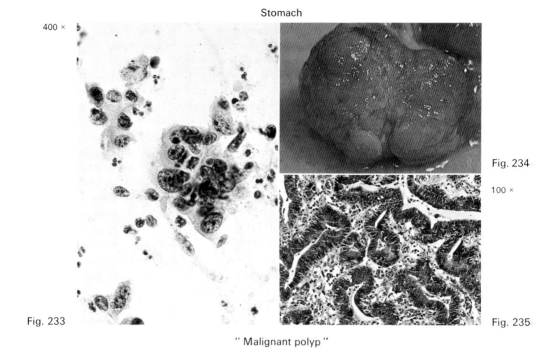

400 ×

Fig. 233

Fig. 234

100 ×

Fig. 235

"Malignant polyp"

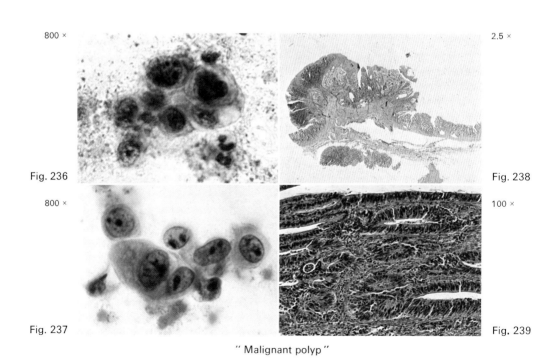

800 ×

2.5 ×

Fig. 236

Fig. 238

800 ×

100 ×

Fig. 237

Fig. 239

"Malignant polyp"

Stomach

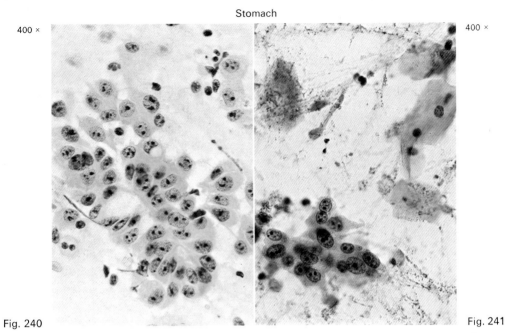

Fig. 240 400× Fig. 241 400×
Well differentiated papillary adenocarcinoma

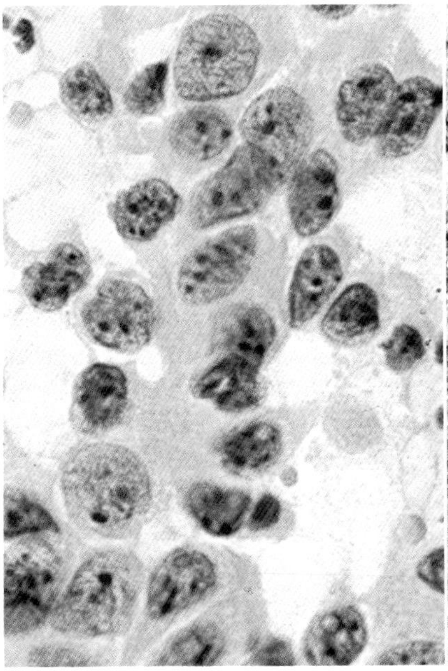

Fig. 242 800× Fig. 243 250×
Well differentiated papillary adenocarcinoma

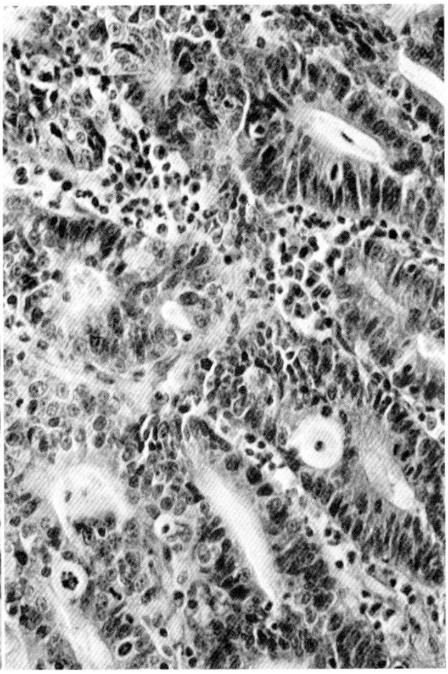

Stomach

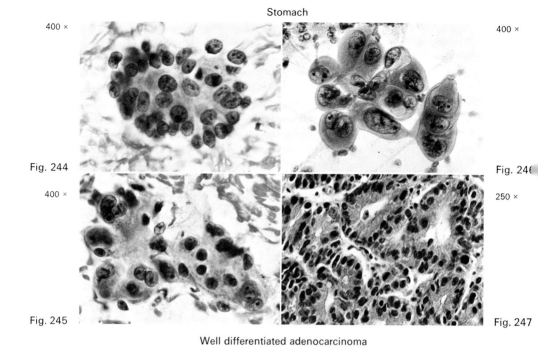

Fig. 244 400×
Fig. 246 400×
Fig. 245 400×
Fig. 247 250×

Well differentiated adenocarcinoma

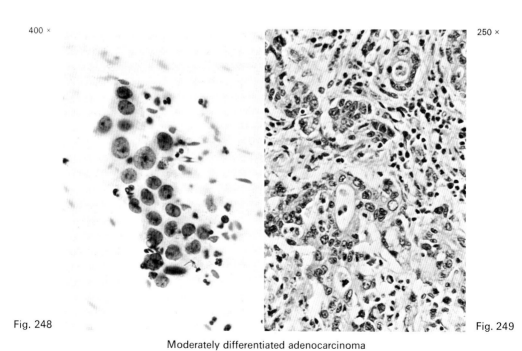

Fig. 248 400×
Fig. 249 250×

Moderately differentiated adenocarcinoma

Stomach

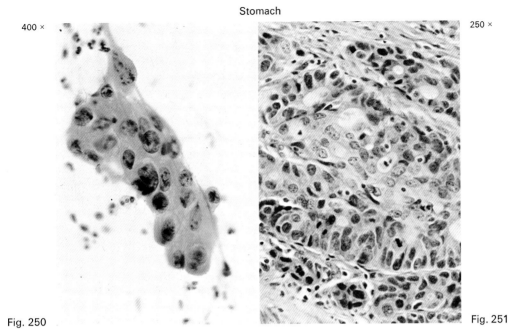

Moderately differentiated adenocarcinoma

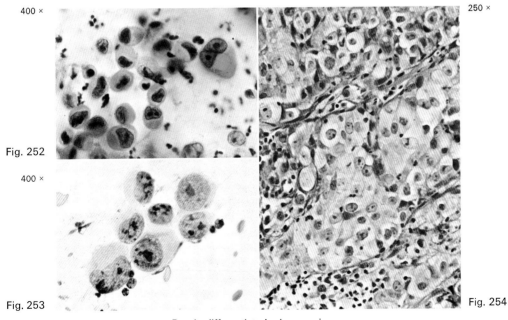

Poorly differentiated adenocarcinoma

Stomach

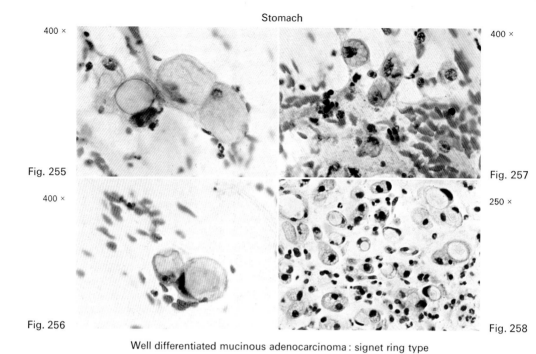

Fig. 255
Fig. 256
Fig. 257
Fig. 258

Well differentiated mucinous adenocarcinoma : signet ring type

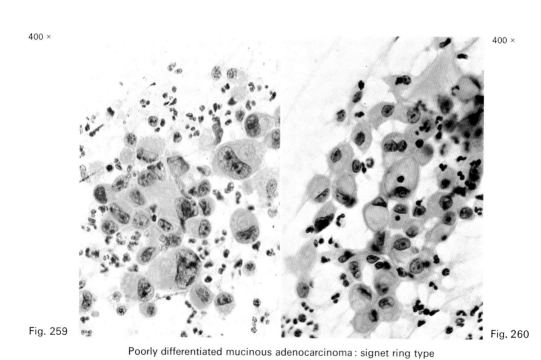

Fig. 259
Fig. 260

Poorly differentiated mucinous adenocarcinoma : signet ring type

Stomach

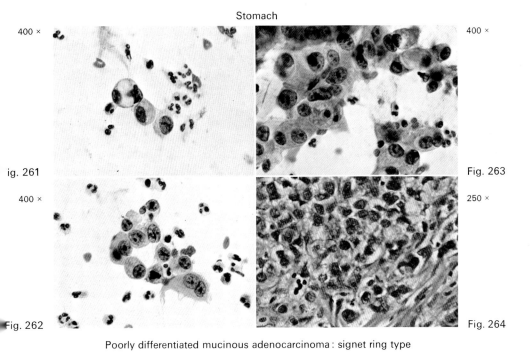

Fig. 261
Fig. 262
Fig. 263
Fig. 264

Poorly differentiated mucinous adenocarcinoma: signet ring type

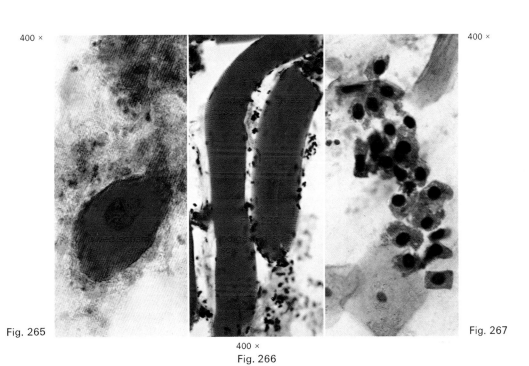

Fig. 265
Swallowed squamous cancer cell in gastric wash from pharyngeal squamous cell carcinoma

Fig. 266
Undigested striated muscle fibre in gastric wash

Fig. 267
Vegetable cells in gastric wash

Stomach

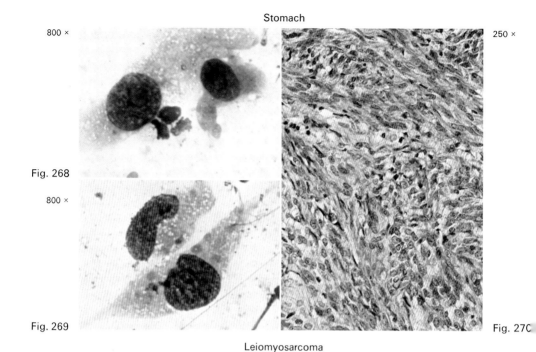

Fig. 268
Fig. 269
Fig. 270

Leiomyosarcoma

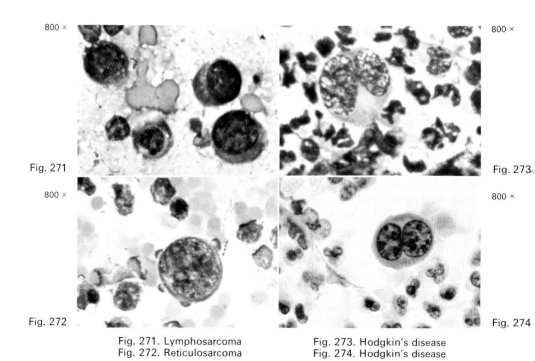

Fig. 271
Fig. 272
Fig. 273
Fig. 274

Fig. 271. Lymphosarcoma
Fig. 272. Reticulosarcoma
Fig. 273. Hodgkin's disease
Fig. 274. Hodgkin's disease

All cytology are May-Grünwald-Giemsa stain except Fig. 274

Thyroid

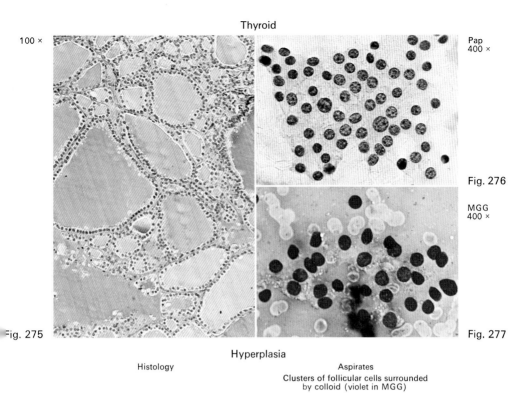

Fig. 275 — 100× — Histology

Fig. 276 — Pap 400×

Fig. 277 — MGG 400×

Hyperplasia

Aspirates
Clusters of follicular cells surrounded by colloid (violet in MGG)

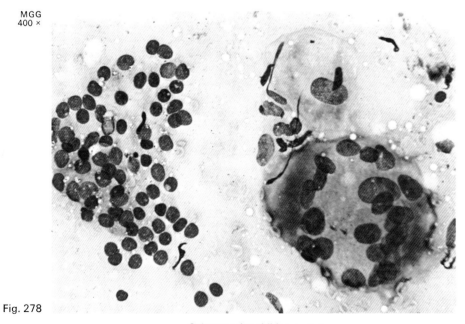

Fig. 278 — MGG 400×

Subacute thyroiditis: Aspirate
A multinucleated giant cell adjacent to a cluster of follicular cells

Thyroid

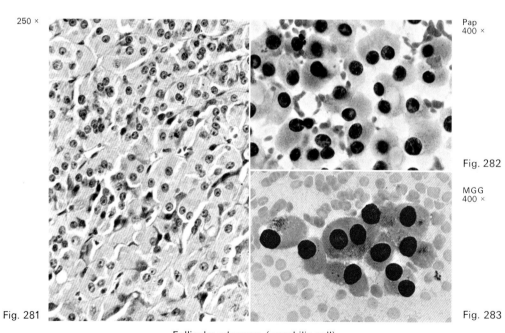

Pap 400 ×
MGG 400 × 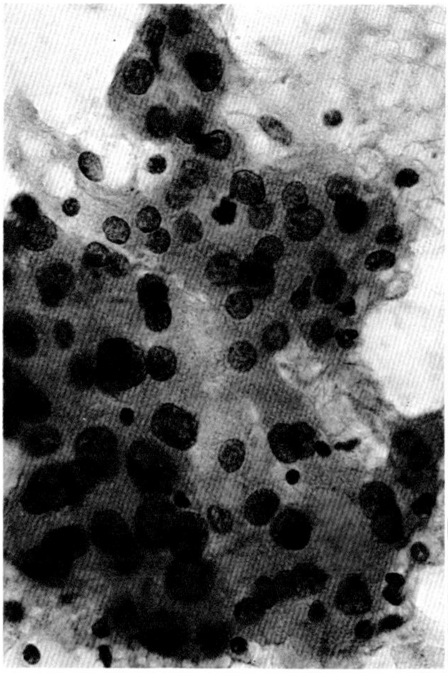
Fig. 279
Fig. 280

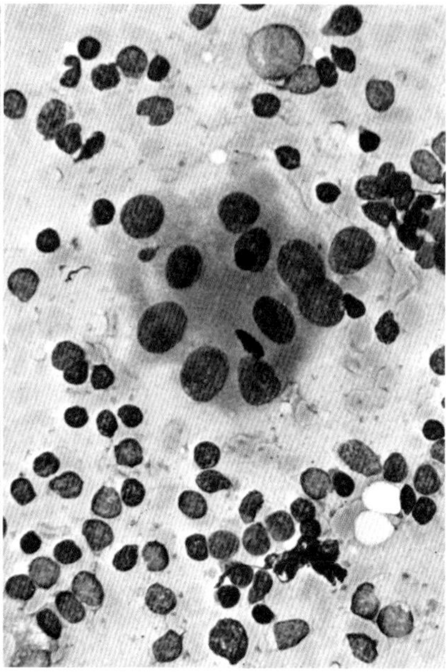

Chronic lymphoid thyroiditis: Aspirates
Enlarged follicular cells with granular cytoplasm and numerous lymphocytes in MGG

250 ×
Pap 400 ×
Fig. 282
MGG 400 ×
Fig. 281
Fig. 283

Follicular adenoma (oxyphilic cell)
Histology   Aspirates
Neoplastic cells with well defined borders
and granular cytoplasm

Thyroid

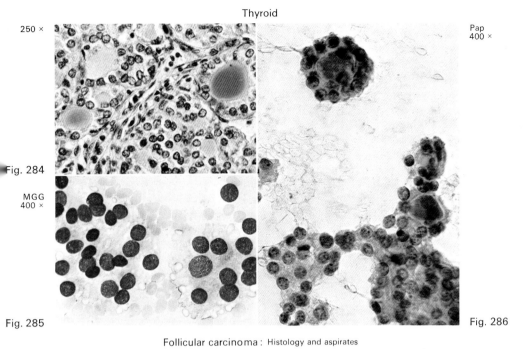

Follicular carcinoma: Histology and aspirates
Colloid is present in two of the follicles

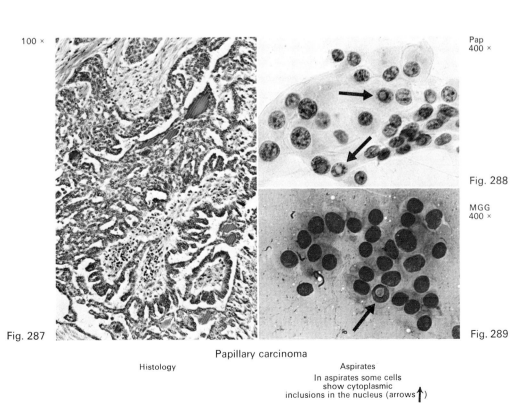

Papillary carcinoma

Histology

Aspirates
In aspirates some cells show cytoplasmic inclusions in the nucleus (arrows ↑)

Thyroid

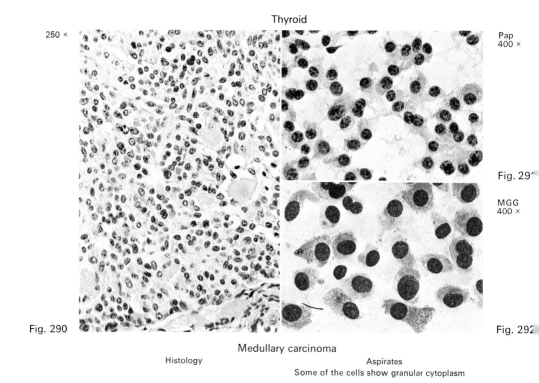

Fig. 290 — Fig. 291 — Fig. 292

Medullary carcinoma
Histology — Aspirates
Some of the cells show granular cytoplasm

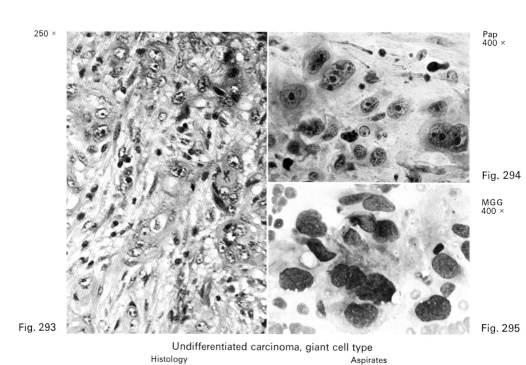

Fig. 293 — Fig. 294 — Fig. 295

Undifferentiated carcinoma, giant cell type
Histology — Aspirates

Breast

100 ×
Pap
160 ×

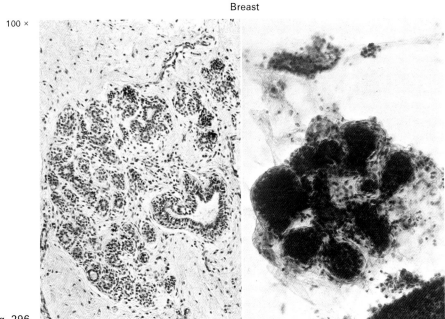

Fig. 296　　　　　　　　　　　　　　　　　　　　　　　　　　　　Fig. 297

Lobular structures

Histology　　　　　　　　　　　　　　　　Aspirate
Lobule with adjacent duct　　　　　　　Intralobular ductules and acini

Pap
400 ×
MGG
400 ×

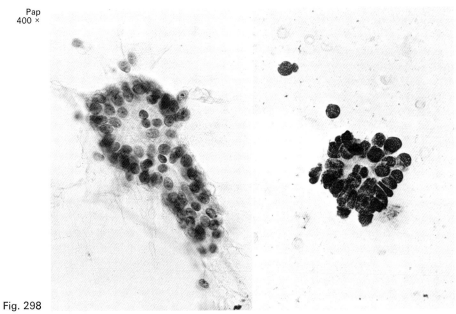

Fig. 298　　　　　　　　　　　　　　　　　　　　　　　　　　　　Fig. 299

Ductular-acinar structures: Aspirates

Breast

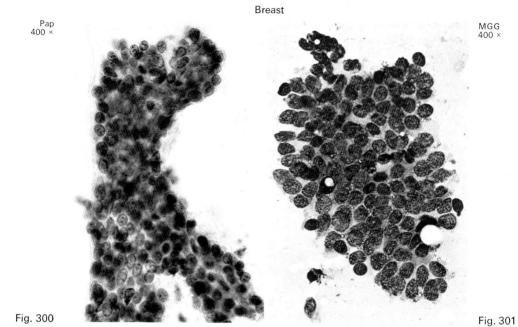

Pap 400 ×
MGG 400 ×
Fig. 300
Fig. 301

Ductal structures: Aspirates

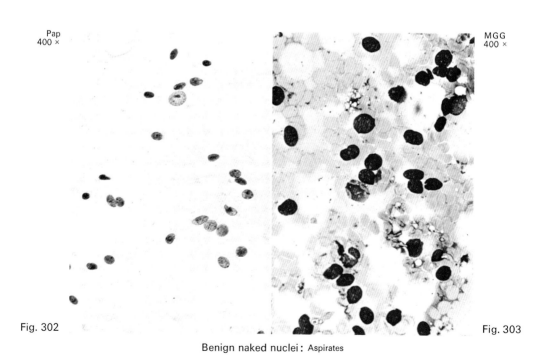

Pap 400 ×
MGG 400 ×
Fig. 302
Fig. 303

Benign naked nuclei: Aspirates

Breast

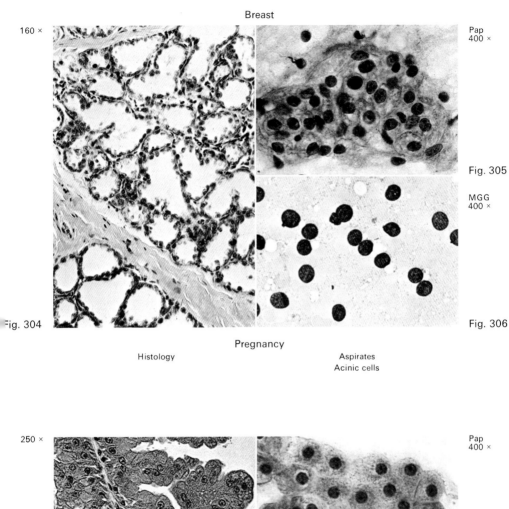

160 ×
Pap 400 ×
Fig. 305
MGG 400 ×
Fig. 304
Fig. 306

Pregnancy

Histology   Aspirates
Acinic cells

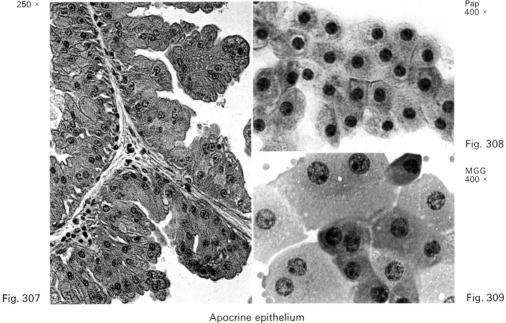

250 ×
Pap 400 ×
Fig. 308
MGG 400 ×
Fig. 307
Fig. 309

Apocrine epithelium

Histology   Aspirates

Breast

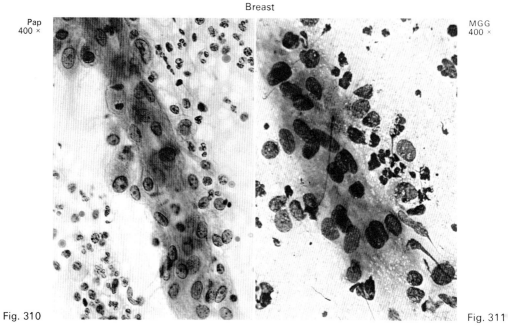

Fig. 310 — Pap 400×
Fig. 311 — MGG 400×

Inflammation: Aspirates
Capillary endothelium surrounded by inflammatory cells

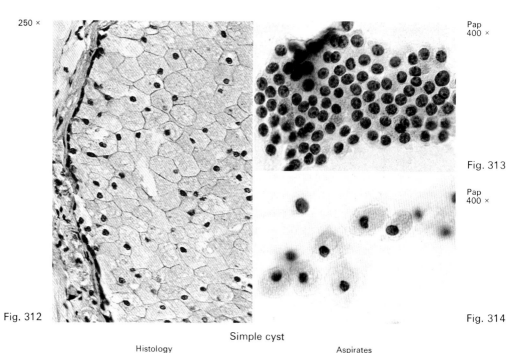

Fig. 312 — 250×
Fig. 313 — Pap 400×
Fig. 314 — Pap 400×

Simple cyst

Histology
Duct: foam cells

Aspirates
Fig. 313. A cluster of lining cells
Fig. 314. Foam cells

Breast

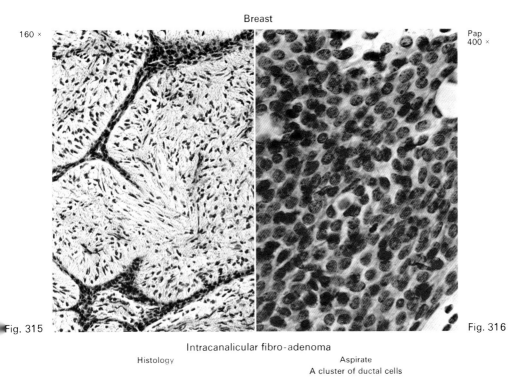

Fig. 315 — 160× — Histology
Fig. 316 — Pap 400× — Aspirate

Intracanalicular fibro-adenoma

A cluster of ductal cells

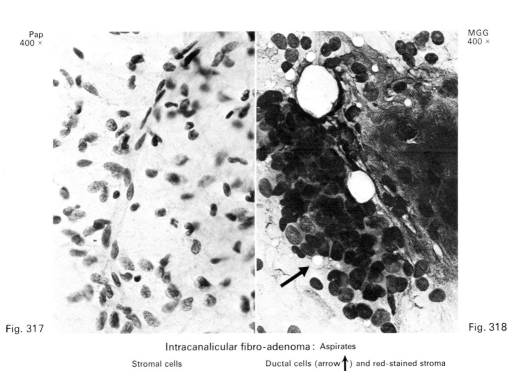

Fig. 317 — Pap 400×
Fig. 318 — MGG 400×

Intracanalicular fibro-adenoma: Aspirates

Stromal cells   Ductal cells (arrow ↑) and red-stained stroma

Breast

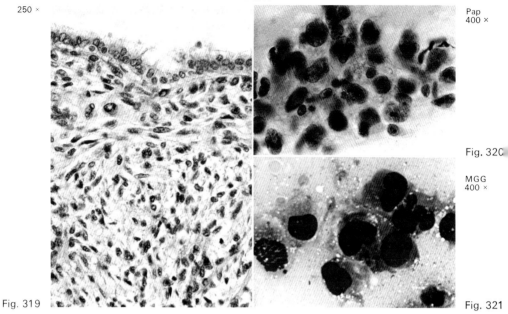

Fig. 319

Fig. 320

Fig. 321

Sarcoma arising in cellular intracanalicular fibro-adenoma
Histology　　　　　　　　　　　　　Aspirates
Benign epithelium and underlying sarcoma　　　Sarcoma cells

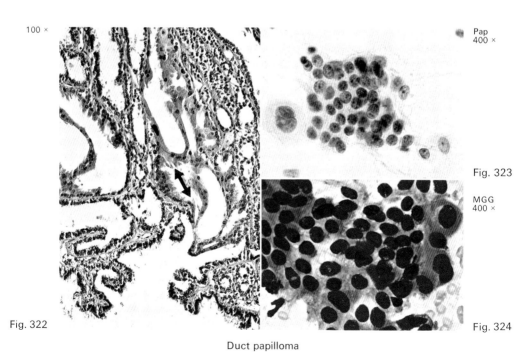

Fig. 322

Fig. 323

Fig. 324

Duct papilloma
Histology　　　　　　　　　　　　　Aspirates
Apocrine metaplasia (arrow↑)　　　Occasional large cells are a characteristic feature

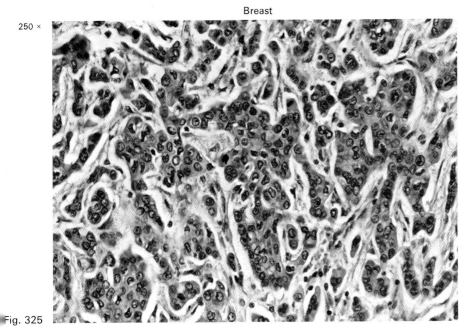

Duct carcinoma: Histology

Duct carcinoma: Aspirates
Clusters resembling ductal structures

Breast

250 ×

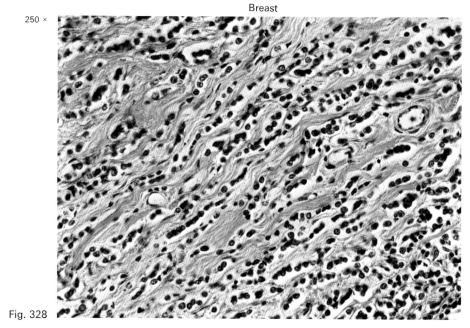

Fig. 328

Lobular carcinoma: Histology

Pap
400 ×

MGG
400 ×

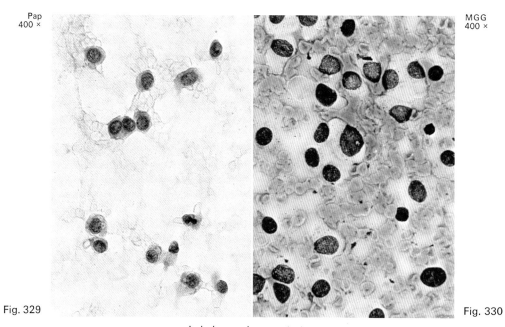

Fig. 329

Fig. 330

Lobular carcinoma: Aspirates
Scanty individual carcinoma cells

Breast

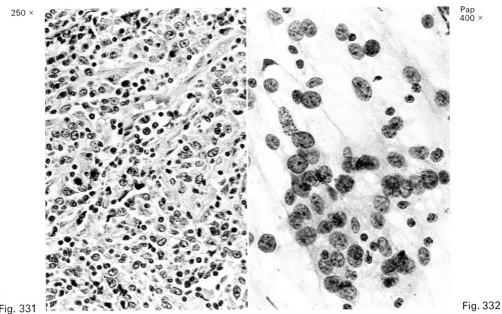

Fig. 331 — 250× Histology
Fig. 332 — Pap 400× Aspirate

Medullary carcinoma

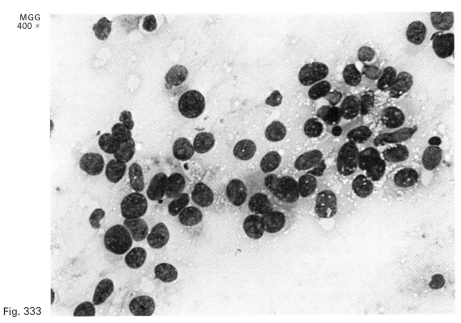

Fig. 333 — MGG 400×

Medullary carcinoma: Aspirate
Indistinct cytoplasmic borders

Breast

250 ×

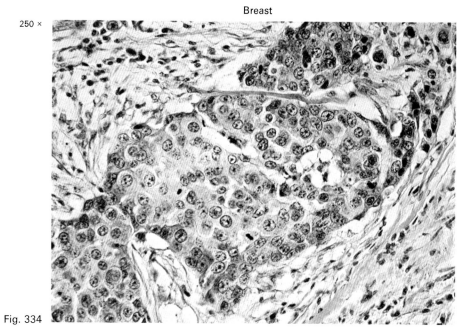

Fig. 334

Apocrine cell carcinoma: Histology

Pap 400 × MGG 400 ×

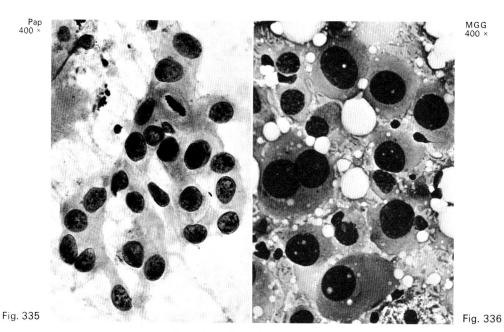

Fig. 335 Fig. 336

Apocrine cell carcinoma: Aspirates
Well defined cell borders

Breast

100 ×
Pap 400 ×

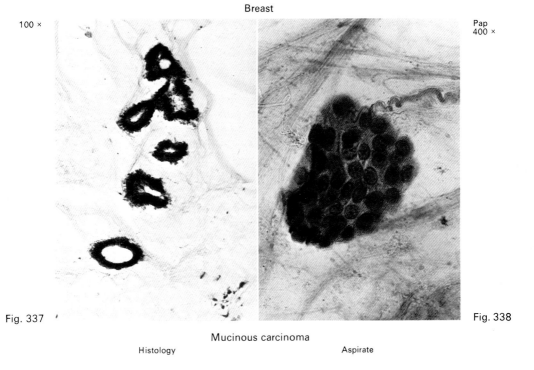

Fig. 337 Fig. 338

Mucinous carcinoma

Histology Aspirate

MGG 400 ×
Alcian blue stain 400 ×

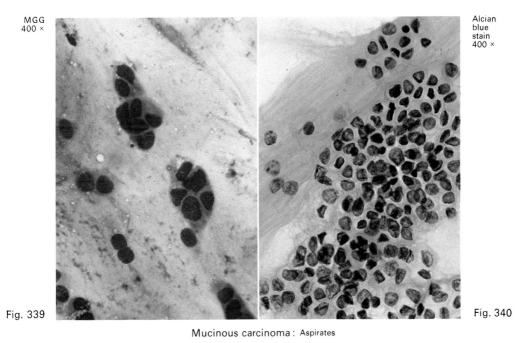

Fig. 339 Fig. 340

Mucinous carcinoma: Aspirates

Prostate

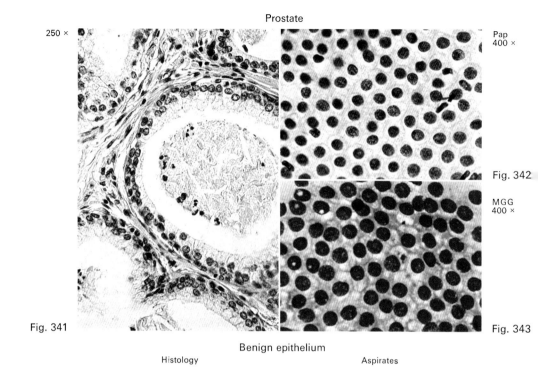

Fig. 341 — 250×
Fig. 342 — Pap 400×
Fig. 343 — MGG 400×

Benign epithelium

Histology   Aspirates

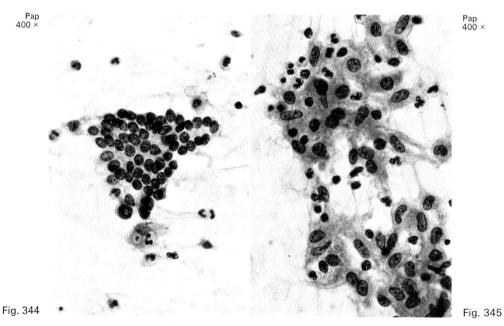

Fig. 344 — Pap 400×
Fig. 345 — Pap 400×

Granulomatous prostatitis: Aspirates

Benign acinar structures surrounded by inflammatory cells   Clusters of histiocytes

Prostate

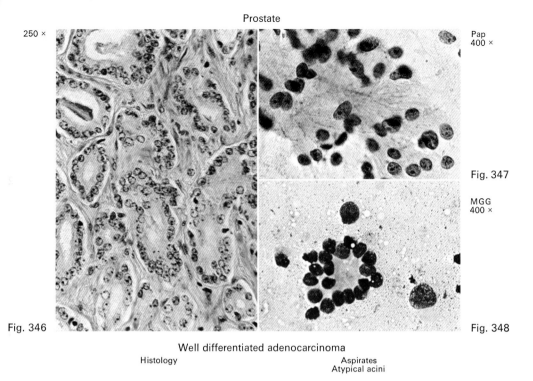

Fig. 346 — 250× — Histology
Fig. 347 — Pap 400×
Fig. 348 — MGG 400×

Well differentiated adenocarcinoma
Aspirates
Atypical acini

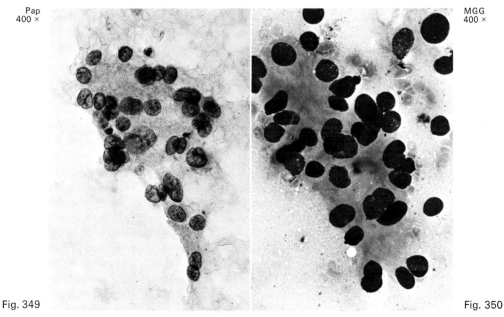

Fig. 349 — Pap 400×
Fig. 350 — MGG 400×

Moderately differentiated adenocarcinoma: Aspirates
Polymorphic carcinoma cells in acinar arrangement

Prostate

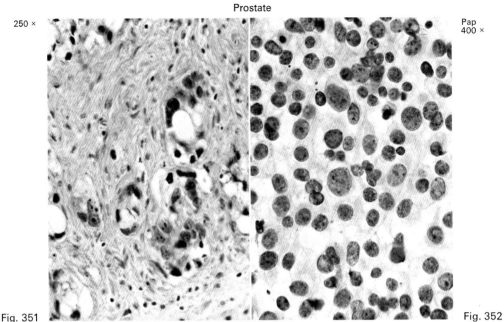

Fig. 351 — 250× Histology
Fig. 352 — Pap 400× Aspirate

Poorly differentiated adenocarcinoma

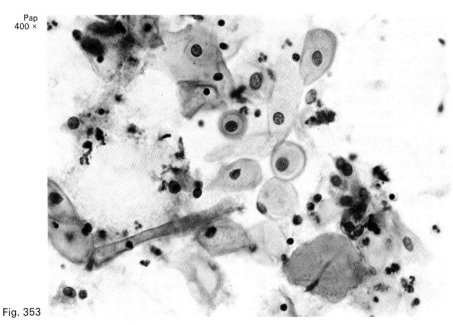

Fig. 353 — Pap 400×

**Well differentiated adenocarcinoma following estrogen treatment:
Metaplastic squamous cells**
Aspirate

Seminal vesicles

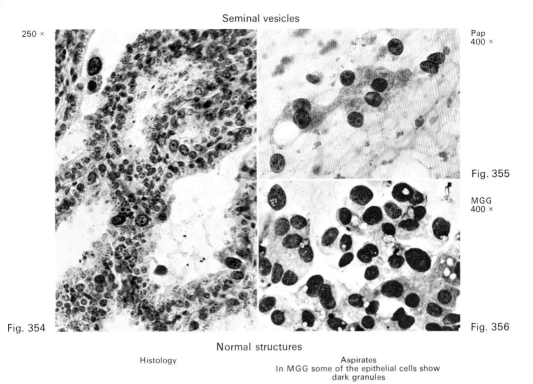

Fig. 354 — 250× — Histology
Fig. 355 — Pap 400×
Fig. 356 — MGG 400×

Normal structures

Histology | Aspirates
In MGG some of the epithelial cells show dark granules

Urinary bladder

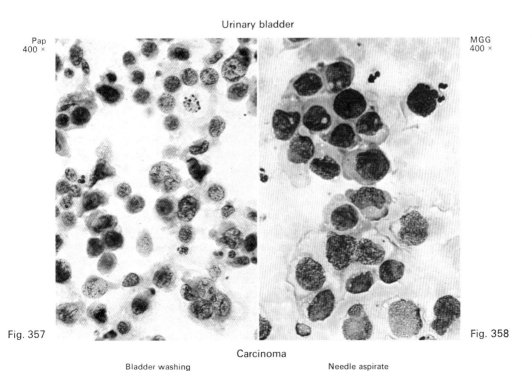

Fig. 357 — Pap 400×
Fig. 358 — MGG 400×

Carcinoma

Bladder washing | Needle aspirate

Lymph node

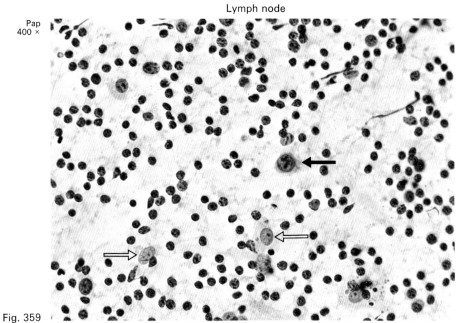

Fig. 359

Lymphoid hyperplasia: Aspirate

Histiocytes (arrow ⇨) and a basophilic blast cell (arrow ➔)

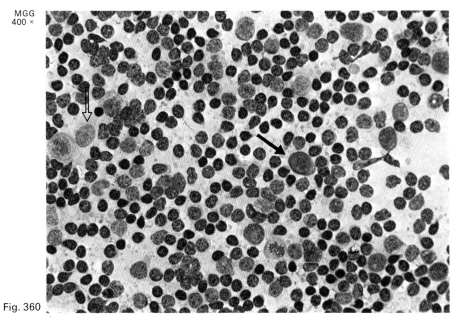

Fig. 360

Hyperplasia: Aspirate

A histiocyte (arrow ⇨) and a basophilic blast cell (arrow ➔)

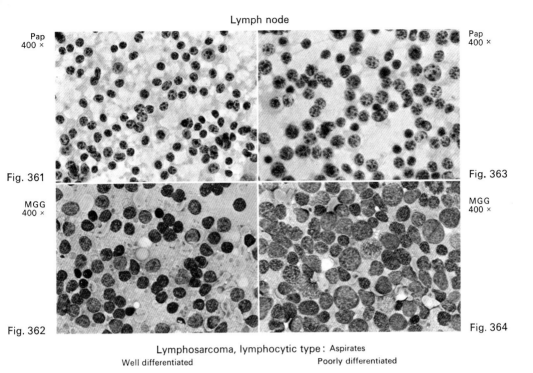

Lymphosarcoma, lymphocytic type: Aspirates
Well differentiated — Poorly differentiated

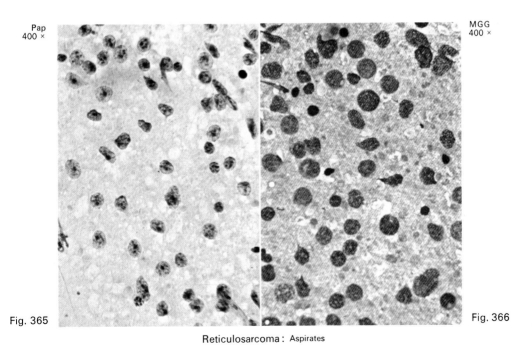

Reticulosarcoma: Aspirates

Lymph node

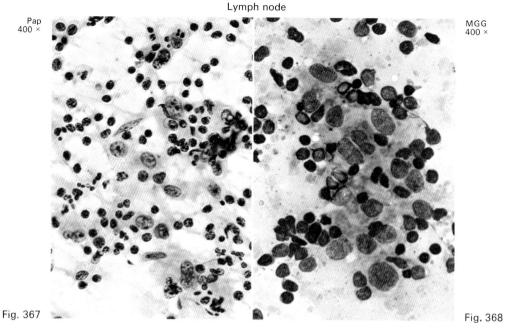

Pap 400 ×
MGG 400 ×

Fig. 367
Fig. 368

Hodgkin's disease: Aspirates
Atypical histiocytes

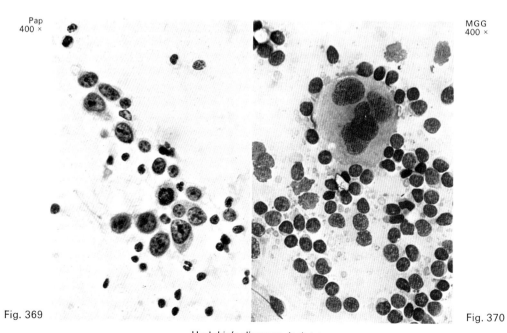

Pap 400 ×
MGG 400 ×

Fig. 369
Fig. 370

Hodgkin's disease: Aspirates
Atypical mononucleated cells
Reed-Sternberg cell

Lymph node

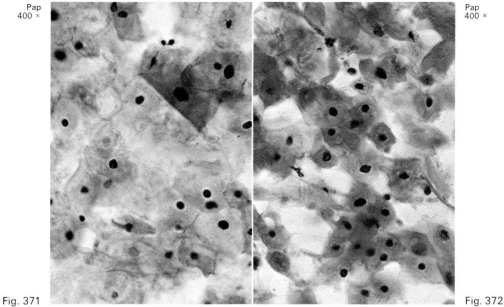

Pap 400 ×
Pap 400 ×
Fig. 371
Fig. 372

Bronchiogenic cleft cyst Lymph-node metastasis of a well differentiated squamous cell carcinoma of the lip

Aspirates
A detailed examination of the smear is required for differential diagnosis

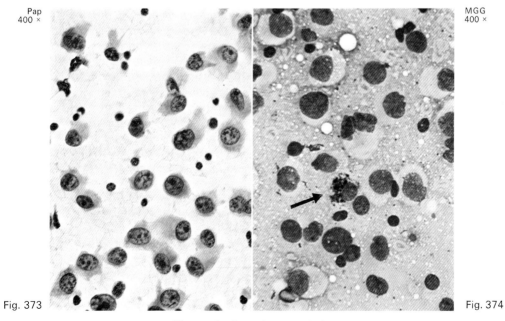

Pap 400 ×
MGG 400 ×
Fig. 373
Fig. 374

Metastasis of a melanoma: Aspirates
Melanin granules (arrow ↑)

Lymph node

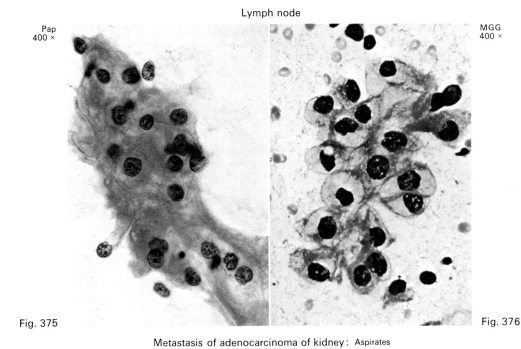

Pap 400 ×

MGG 400 ×

Fig. 375

Fig. 376

Metastasis of adenocarcinoma of kidney: Aspirates

Pap 400 ×

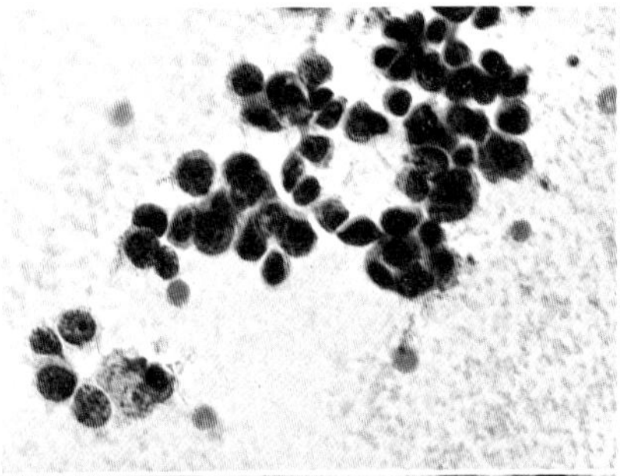

Pap 400 ×

Fig. 377

MGG 400 ×

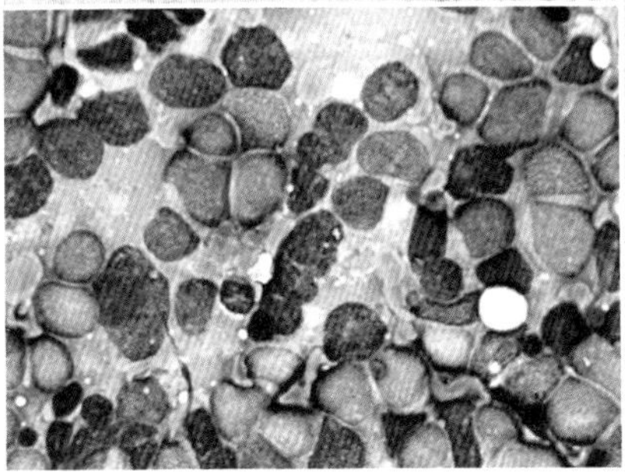

Fig. 378

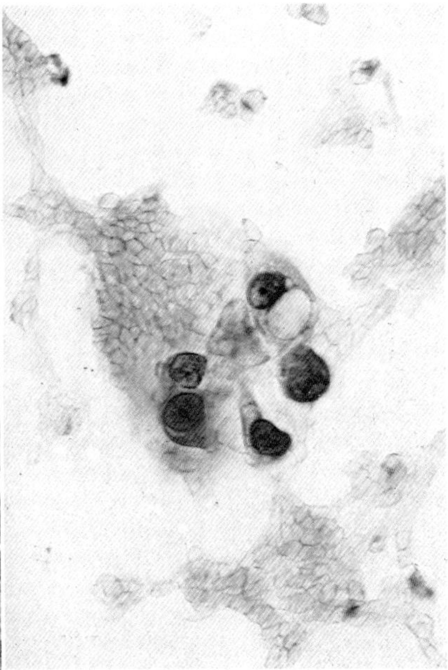

Fig. 379

Metastasis of oat cell carcinoma of lung   Metastasis of mucinous (signet ring) carcinoma of stomach

Aspirates